Domestic Satellite: An FCC Giant Step

Westview Replica Editions

This book is a Westview Replica Edition. The concept of Replica Editions is a response to the crisis in academic and informational publishing. Library budgets for books have been severely curtailed; economic pressures on the university presses and the few private publishing companies primarily interested in scholarly manuscripts have severely limited the capacity of the industry to properly serve the academic and research communities. Many manuscripts dealing with important subjects, often representing the highest level of scholarship, are today not economically viable publishing projects. Or, if they are accepted for publication, they are often subject to lead times ranging from one to three years. Scholars are understandably frustrated when they realize that their first-class research cannot be published within a reasonable time frame, if at all.

Westview Replica Editions seem to us one feasible and practical solution to the crisis. The concept is simple. We accept a manuscript in camera-ready form and move it immediately into the production process. The responsibility for textual and copy editing lies with the author or sponsoring organization. If necessary we will advise the author on proper preparation of footnotes and bibliography. The manuscript is acceptable as typed for a thesis or dissertation or prepared in any other clearly organized and readable way, though we prefer it typed according to our specifications. The end result is a book produced by lithography and bound in hard covers. Edition sizes range from 200 to 600 copies. We will include among Westview Replica Editions only works of outstanding scholarly quality or of great informational value and we will exercise our usual editorial standards and quality control.

Domestic Satellite: An FCC Giant Step
Toward Competitive
Telecommunications Policy

Robert S. Magnant

In a society where as much as 50 percent of the gross national product is spent on information transfer services, trends in communications policy and capabilities become extremely important and can have major consequences. During the past ten years, the Federal Communications Commission has made a major effort to promote competition and innovation in the telecommunications industry. But, in order that public interests be identified and adequately served, there must be a wider understanding of the industry and the issues. Robert Magnant takes full account of the multidisciplinary nature of the telecommunications field, and in this book presents an objective treatment of technological trends, policy issues, and market constraints. His study has been designed to provide intelligible, easily accessible information for the legislator, the business leader, the concerned citizen. It deals with the specific area of telecommunications, but in its complete coverage of a major domestic policy issue, it is at the same time an excellent case study to illuminate the intricacies and options in the formulation and operation both of general government and business policy and, more specifically, regulated market structures.

Robert S. Magnant is chief engineer and technical director for the U.S. Army Communications Electronics Engineering Installation Agency–Continental United States (USACEEIA-CONUS). In 1976, while on academic leave from his position with USACEEIA, he received a master's degree in telecommunications from the University of Colorado.

Domestic Satellite: An FCC Giant Step

Toward Competitive Telecommunications Policy

Robert S. Magnant

Routledge
Taylor & Francis Group
LONDON AND NEW YORK

First published 1977 by Westview Press

Published 2018 by Routledge
52 Vanderbilt Avenue, New York, NY 10017
2 Park Square, Milton Park, Abingdon, Oxon OX14 4RN

Routledge is an imprint of the Taylor & Francis Group, an informa business

Library of Congress Cataloging in Publication Data

Magnant, Robert S.
Domestic satellite.

(A Westview replica edition)
Bibliography: p.
1. United States. Federal Telecommunications Commission.
2. Artificial satellites in telecommunication. I. Title.
HE9721.U5M33 1977 384.54'56 76-30840

ISBN 13: 978-0-367-01695-1 (hbk)

To the memory of Eric W. Muller, Jr. --

a true professional and a very dear friend.

TABLE OF CONTENTS

LIST OF TABLES. .xii

LIST OF ILLUSTRATIONS xiii

PREFACE xv

ACKNOWLEDGEMENTS. xvii

INTRODUCTION. 1

Chapter

I. HISTORICAL FOUNDATIONS AND BOUNDARY CONDITIONS 9

A. The Federal Communications Commission and the Common Carrier Industry 10

1. Regulatory Origins. 11

a. Common Carriers Defined 15
b. Domestic Common Carrier History . 17

2. The Federal Communications Commission. 19

a. The Public Interest 21
b. The Regulated Common Carriers . . 24

B. The Constraints of Business and Technology. 28

1. The Market Structure. 29

a. Market Economics. 31
b. Rate of Return Regulation 35
c. Competition and Antitrust 39

2. Satellite Technology. 46

a. Orbit Considerations. 49
b. Frequency Allocations 51

TABLE OF CONTENTS (continued)

II. THE PRECEDENTS OF DOMSAT. 57

A. The Communications Satellite Act of 1962. 58

1. Congressional Hearings. 59

a. The Ownership Question. 60
b. The Kennedy Administration Bill . 64

2. H.R. 11040 Becomes Law. 67

a. Opposing Views. 68
b. The Final Act 70

B. Space Age Regulation. 74

1. International Considerations. 76

a. The International Telecommunication Union. 77
b. Orbital Parking Slots 83

2. Authorized Users. 88

3. The DOMSAT Question 91

III. THE DEVELOPMENT OF THE DOMSAT POLICY. 97

A. The Basic Issues. 99

1. A Notice of Inquiry 100

a. The Ford Proposal 102
b. A Question of Law 108

2. Technology. 112

a. Technology Dynamics 114
b. Computercations 118
c. The Computer Inquiry. 122

3. The "New Competition" 131

a. Interconnection 133
b. The Specialized Carriers. 136

TABLE OF CONTENTS (continued)

B. The Policy Makers 140

1. Activities of the Johnson Administration 143

a. The Johnson Task Force. 145
b. The National Academy of Engineering 149

2. A Change of Administrations 153

a. The Commission. 157
b. The Office of Telecommunications Policy. 161
c. Congressional Oversight 163

3. The Final Proceedings 169

a. Conflicting Opinions. 173
b. Competition Reemphasized. 177
c. Backlash to Competition 183

IV. DOMSAT POLICY TODAY. 193

A. Trends in Technology. 195

1. Data Projections. 197

2. Network Developments. 200

a. SPADE 201
b. ARPANET 204
c. Packet Economics. 208
d. Line Control Procedures 211

3. The ALOHA Concept 214

a. Pure ALOHA. 215
b. Random Packets. 217
c. Satellite Channels. 219

B. The IBM Entry--SBS. 226

1. A Joint Proposal. 228

2. The CML Decision. 232

a. SDLC. 234
b. Commission Findings 239

TABLE OF CONTENTS (continued)

3. The SBS Filing Today. 242

a. The "Bell Bill" 244
b. Potential Competition for AT&T. .248

CONCLUSIONS .255

SELECTED BIBLIOGRAPHY 259

APPENDICES. .275

A. Letters, Interviews, or Documents Furnished .277

B. A Chronology of Individuals Interacting in DOMSAT. 283

C. Further Analysis of the ALOHA Concept 285

LIST OF TABLES

1. Data Traffic Forecast for the United States . .200

2. Common Signalling Channel Performance Parameters 203

3. Signalling and Switching Processor Parameters 205

4. Computer Resource Usage Within ARPANET. 210

5. Cost Estimates for INTELSAT Communication Satellites 221

6. Summary of Major Requirements 236

LIST OF ILLUSTRATIONS

1. Population and Number of Telephones Seen From the Geostationary Satellite Orbit. . . 86

2. The World of the FCC 98

3. Projected Information Transfer Volume, 1970-1990 199

4. ARPANET-March 1972 208

5. The ALOHA System 216

6. The SDLC Frame 237

7. ALOHA Multiplexing 286

8. Channel Utilization vs. Channel Traffic. . . .290

9. Priority Queues. 292

10. Broadcast Channel Flow Control 295

PREFACE

The Federal Communications Commission is the licensing and regulatory authority for the private, nongovernmental interstate and foreign telecommunications of the United States. Consequently its activities affect virtually everyone either directly or indirectly. Since its establishment by Congressional legislation in 1934, its activities, its personalities and its policies have occasionally been the object of lengthy discussion and debate; however, its role during the past ten years has been particularly controversial.

Instead of merely interpreting existing law, the Commission, by necessity, must continually break new ground in order to accommodate new uses of communications technology as was exemplified by its decisions in the applicaticn of satellites to domestic communications and the interdependence of computers and communications. In addition, since 1959, the Commission has promoted a definite policy of increased competition in the industry. Such groundbreaking efforts are now manifesting themselves in the current filing before the Commission of Satellite Business Systems [an IBM sponsored partnership] for a wideband, digitized domestic satellite system.

A review of the factors that have influenced the Commission's domestic satellite policy provides a picture of the increasingly complex world of this Federal regulatory agency. The main hypothesis of this study is that this policy was a profound decision by the Commission [possibly the most important and most complex in the Commission's history] with a lasting potential for benefit to the public inherent in it. The significant points of technology and policy raised by the SBS filing are discussed to show the potential of SBS to offer innovative services to data communications users and to compete with the communication industry's dominant monopoly, American Telephone and Telegraph.

This subject is considered to be an excellent example of interdisciplinary interactions in the field of telecommunications. An appreciation for such interactions is generally necessary when addressing any telecommunication issue. It is intended that this study will give added emphasis to the merging of computer and communications technologies and promote an increased appreciation for satellite regulation at both the national and international levels.

ACKNOWLEDGMENTS

The following pages are dedicated to Professionals; people of the highest standards--like Geoffrey Ballard, Les Callahan, Irwin Higgs, Bill Kenneally, W. B. Latta, Norman Shupe, and the members of the U.S. Army's Communications-Electronics Engineering Installation Agency--as well as many others whom I have either known or worked with within the Department of the Army over the past twelve years. These are motivated, hard-working people who believe in what they are doing. There are such people within the Federal Communications Commission.

Sy Greenspan and Harold Randall deserve particular note for having done more than their fair share; guiding me, teaching me (with much patience I might add), believing in me, and consequently making this year and this work possible. The support of my wife, Rosemary, goes without saying.

Special appreciation is extended to Dr. George A. Codding, Jr., Dr. Frank S. Barnes and the University of Colorado faculty for their guidance and assistance in the development of this effort. Jim Alleman and the members of the Policy Research Division staff at the Department of Commerce, Office of Telecommunications, have contributed valuable commentary to me throughout this study and I am particularly indebted to Cheryl Zegers, Technical Information Specialist at the Department of Commerce's Boulder Library, for her data-gathering assistance.

The information and support contributed to this effort by the individuals listed in Appendix A through interviews and/or correspondence and documentation will never be fully quantified. I wish to also thank Dick Talley of Hughes Aircraft for his critique of the draft manuscript and Dale Hatfield and Seb Lasher of the Federal Communications Commission for their comments and criticisms of this effort.

Any viewpoints found in the text without citations are those of the author and should not be construed as the opinions or the official interpretations of policy of the individuals or organizations discussed.

R.S.M.

Boulder, Colorado
July, 1976

INTRODUCTION

In deference to Arthur Clark and Marshall McLuhan, this study begins with neither a quote from "Extraterrestrial Relays" nor a profound discussion of the satellite's molding of the human race into a global village.[1] The existence of communications satellites is as second nature to the bicentennial American as moon landings. The "LIVE VIA SATELLITE" caption at the bottom of a television picture (if it's even indicated anymore) no longer gives special cause for notice or excitement. Today, without having direct involvement in either the aerospace or telecommunications industry, the average individual could easily conclude that the "communications by satellite" revolution was a product of the 60's and think no more of it. In fact, satellite technology has had and continues to have a major impact on the capabilities of today's world-wide telecommunication systems. Its applications to domestic communications are just beginning to be developed and several options

---

[1]Clark's 1945 article, predicting the use of synchronous satellites for communication, appears in the Appendix of his Voices from the Sky (New York: Harper and Row, 1965). Also see Neil P. Hurley, "Marshall McLuhan: Communications Explorer," The McLuhan Explosion, H. H. Crosby and G. R. Bond, eds. (New York: The American Book Company, 1968), pp. 154-59.

are either in the proposal stage or being implemented.

Telecommunications has been defined by the International Telecommunication Union as:

> Any transmission, emission or reception of signs, signals, writing images and sound or intelligence of any nature by wire, radio, optical, or other electromagnetic systems.

Unfortunately, there are many people who are unaware of this definition, or any simpler definition for that matter, of what telecommunications is and what it means to their individual lives. Complicating any understanding of telecommunications are the economic considerations of regulated and competitive markets. The United States, traditionally a competitive, free-enterprise market, is one of the few countries of the world where telecommunications is part of the private sector, as opposed to national systems. Private concerns furnish communications services to the nation that are "affected with a public interest." Consequently these firms are designated as "public utilities" and from an economic standpoint, they possess technological characteristics that almost inevitably result in monopoly market structures. American Telephone and Telegraph (AT&T) has long dominated this nation's communications market.

Generally, it is agreed that where the common benefit is dominant, where the whole of society is involved, economic functions will be performed by society itself. Further, where the common interest requires interference with private functions, government will intervene. Public utilities and similar regulated industries are a "halfway house" between completely government functions and free enterprise functions.[2] Regulation is imposed by the government to fix reasonable prices for the services rendered as a substitute for competition. Under the present structure, the Federal Communications Commission (FCC) is the government body having primary responsibility for regulatory policy in the telecommunications area.

In domestic telecommunications, a changing market structure and a pervasive rate of technical innovation have fostered a dynamic regulatory environment. Since 1959 the communications common carrier industry has been undergoing a transition as a result of several new policies that have been instituted by the FCC to promote competition in the industry. Also the rapid rate of technological innovation of the 1960's has blended, if not merged, the computer and communications technologies together. One of the Commission's more recent policies, Domestic Satellite, stands out as unique and seems to be the embodiment of all the pro-competitive policies of the FCC to date.

---

[2]Martin T. Farris and Roy J. Sampson, Public Utilities: Regulation, Management and Ownership (Boston, Massachusetts: Houghton Mifflin Company, 1973), pp. 6-8.

The fact that the common carrier industry is a traditionally regulated industry makes this policy and the Commission's role as regulator only more important.

It has been suggested by Adams and Dirlam that nothing could better illustrate the pressures that a regulatory commission must resist [in the execution of its duties during periods of dynamic technological change] than the satellite.[3] On March 2, 1966 the Commission formally initiated a Notice of Inquiry, Docket No. 16495--In the Matter of Establishment of Domestic Noncommon Carrier Communications Satellite Facilities by Nongovernmental Entities, but it was over six years later before it finalized a "limited open entry" policy for domestic communications satellites. Through a review of official documents, literature searches, formal correspondence and personal interviews, this research examines the factors which appear to have influenced the Commission's Domestic Satellite [also to be referred to as DOMSAT] proceedings and identifies present considerations that have been placed before the Commission since that ruling. Such analysis:

(1) provides a comprehensive picture of the multi-faceted interface that the FCC has with its

---

[3]Walter Adams, Joel B. Dirlam, "Market Structure, Regulation, and Dynamic Change," Performance Under Regulation, Harry M. Trebing, ed. (East Lansing, Michigan: MSU Public Utilities Studies, 1968), pp. 131-144.

environment,

(2) demonstrates how interrelated the issues can become when determining policy in an area of dynamic technological change,

(3) shows how the inherent technical characteristics of communications satellites [which have no exact terrestrial equivalent] and the advances in computer-communications have contributed to the complexity of this issue, and

(4) identifies instances where the satellite policies of four different presidential administrations, compounded with an assortment of study groups, personalities, industry postures and international considerations confounded the issue before the Commission.

The objective of this study is that it serve as a vehicle for increasing the "public's awareness" to the subject of telecommunications and to the status of the domestic satellite issue and, as a consequence, lend support to the Commission in its current and future efforts. The FCC's performance in the regulation of today's common carriers has not been receiving the respect it deserves. However, confidence in the Commission's capabilities and the effectiveness of the regulatory process is central to the public and national interest.[4]

---

[4]Final Report, President's Task Force on Communications Policy, December 7, 1968 (Washington, D.C.: U.S. Government Printing Office, 1969), 0-351-636, Tab B, p. 9.

The FCC comprises men and women, professionals in their fields, who are attempting to perform an enormous task with limited resources. It is impossible for them to have all the right answers all the time in such a complex world as theirs. Even Sir Arthur Clark, looking back on the proposition of patenting his 1945 concept, notes:

> The idea of patenting the geostationary communications satellite concept never occurred to me and my excuse for this is sheer lack of imagination.[5]

This study offers a positive perspective of Commission's efforts in an area of dynamic technology. The adequacy of the FCC's organizational structure is not an issue but rather a factor which is addressed in passing only.

The analysis of the Commission's domestic satellite considerations is arranged to follow the historical pattern of events surrounding DOMSAT. The problem, however, is initially set within the framework of the business considerations and the technical limitations that existed. The time value of such information played an important role in the policy-making process. Time controls the available technology, it defines the existing and projected business markets and it determines the political

---

[5]Arthur Clark, Voices from the Sky (New York: Harper and Row, 1965), p. 125.

priorities of the day. The emphasis given to each of these factors varied throughout the DOMSAT proceedings.

This study of the policy-making process divides itself into three distinct periods:

(1959-1965) - the precedents of DOMSAT.

(1966-1972) - the development of the DOMSAT policy.

(1973 to present) - DOMSAT policy today.

The logical starting date for this review is 1959 as it was in December of that year that President Eisenhower first spoke of the commercial use of communications satellites. Also 1959 is a well-documented date for the beginning of the FCC's current policy of competition. The initial period of discussion is from 1959 up until the DOMSAT question was raised by the American Broadcasting Company's filing in 1965 and is covered in Chapter II. This period includes the policy precedents of the Eisenhower and Kennedy Administrations that resulted in the Communications Satellite Act of 1962.

Chapter III covers from March 2, 1966, the date of issuance of the Commission's Notice of Inquiry on DOMSAT (Docket No. 16495) through December 22, 1972, the date of the Commission's final Memorandum Opinion and Order. This is the period when formal DOMSAT policy was defined. Chapter IV looks at DOMSAT from then until today, focusing briefly on some of the results of that decision, and more specifically on the activity surrounding the filing by Satellite Business Systems from current business, technical and regulatory perspectives.

Primary information sources used for this study were official FCC Notices, Reports, Orders and Memorandums as well as filings, briefs and comments submitted by the industries involved in DOMSAT. Official Congressional documentation was used to a large extent as was current formal correspondence from the individuals listed in Appendix A. Automated data base searches were also used for this research and the opinions and analyses used in the following discussions are viewpoints taken from the appropriate periods of time to the greatest extent possible. The formal correspondence noted served both as primary sources of information and as guides which provided direction to the research. Secondary sources of information were textbooks, journals, newspapers, presentations, and published reports.

CHAPTER I

HISTORICAL FOUNDATIONS AND BOUNDARY CONDITIONS

To understand the impact of the *Domestic Satellite* decision, an understanding of the origins of regulation, the domestic communication common carriers and the Federal Communications Commission is required. The roots of the FCC date back more than fifty years to the early days of radio. The legislation by which Congress established this independent agency to regulate the nation's communications and encourage the larger and more effective use of radio in the public interest remains essentially unchanged today.

The world of the FCC is far from simple. The Commission must interface with its environment in a multitude of ways in the performance of its regulatory functions and without a doubt, the common carriers dominate this interface. Technology and market considerations have shaped the industry's structure but both vary with time and both have imposed constraints on the policy makers and have limited the alternatives for them.

This chapter reviews the foundations of the

organizations and industries that participated in DOMSAT. These provide the initial conditions (as well as the constraints) from which a new domestic industry was launched with initial annual revenue estimated in excess of one-half billion dollars and initial investment estimates of almost three times that value.[1] Technology was a major consideration throughout the DOMSAT proceedings. Thus definitions of the boundaries which it created will increase the reader's appreciation for the issues that were before the Commission. Working definitions of competition, regulation, the common carriers, the rate base and the public interest are also provided to establish the baselines that business considerations imposed on DOMSAT.

A. The Federal Communications Commission and the Common Carrier Industry

The one hundredth anniversary of the invention of the telephone and the beginning of the communications common carriers is being celebrated this year. In contrast the FCC is only forty-two years old and generally considered to be a "late bloomer"; its effectiveness as a regulator has only been noticeable during the last two decades. Perhaps this is because demands for new and different services surfaced during

---

[1]John McDonald, "Getting Our Communication Satellite Off the Ground," Fortune, 86 (July, 1972), p. 69.

this period as a result of technological advances.

The improvements in the appearance of a modern telephone instrument over an antique device are in no way a measure of the service improvements available. Today's telecommunications systems, when compared against yesterday's predictions, are orders of magnitude greater than the wildest dreams imagined attainable by our ancestors and they extend far beyond the realm of voice communications and the traditional common carriers.

Although the Commission and the common carriers evolved separately, it is important that their origins be understood. These perspectives which include information relative to the Commission's formation, the roots of the Nation's carriers, the concepts of regulation, the related legislation and the basic form of the industry are considered elementary but necessary background for this study. The carriers, both old and new, and the FCC are the primary elements involved in DOMSAT.

1. Regulatory Origins

To operate a broadcasting station in the United States, one must first obtain a license from the Federal Communications Commission. The delivery of a license is not an automatic function but is at the discretion of the Commission; it is theirs to decide.

How the Commission came to exist and how it acquired such power is a story that spans the first third of this century.

Radio was first used commercially for ship-to-shore and ship-to-ship communication. However, as early as 1901 low frequency radio began to be used to provide overseas radio telegraph services.[2] As a means of communication, radio's facilities are limited. Radio transmission, the transfer of messages by electromagnetic radiation through space rather than along wires or cable, makes use of the frequency spectrum, a limited natural resource.[3] Two radio transmission systems may not employ the same frequencies at the same time in the same area without interfering with one another. Thus there is a fixed natural limitation upon the number of stations that can operate without interfering with one another. Prior to World War I, questions of interference arose rarely because there were more than enough frequencies for the existing number of stations and the state of the art.[4]

---

[2]Kurt Borchardt, Structure and Performance of the U.S. Communications Industry (Boston: Harvard University Press, 1970), p. 46.

[3]This is a range of frequencies (rates of oscillation) of electromagnetic radiation. A radio differentiates between two or more electromagnetic signals or frequencies just as the ear differentiates between different frequencies of sound, such as different notes of music or different voices.

[4]Donald M. Gillmor and Jerome A. Barron, Mass Communication Law, second edition (St. Paul, Minnesota: West Publishing Company, 1974), pp. 763-764.

On August 13, 1912, the Radio Act of 1912 received the approval of both the Senate and the House and became law. It provided that anyone operating a radio station must have a license issued by the Secretary of Commerce. The main difference between the Act and previous bills that had been introduced was that specific regulations were now set out in the Act whereas, previously, power to make regulations had been given to the Secretary of Commerce.[5] Although the Act was primarily designed for maritime communication and "safety at sea" was the reason usually cited for its introduction, R. H. Coase notes that public business such as wireless telegraphy was being hindered and that the true intent of the Act was to bring about government control of the operations of the industry as a whole.[6]

The war accelerated the development of radio and the broadcast industry came into being in the early 1920's. By November 1, 1922 there were 564 broadcasting stations in the United States and Mr. Herbert Hoover, as Secretary of Commerce, was responsible for the administration of the 1912 Act.[7] The

---

[5]R. H. Coase, "The Federal Communications Commission," The Journal of Law and Economics, 2 (October, 1959), pp. 2-3.

[6]Ibid. Each radio station considered itself independent and a state of chaos existed in many places where numerous stations tried to communicate simultaneously.

[7]Ibid., p. 5.

first government/industry Radio Conferences were held in 1923, 1924 and 1925 at which recommendations were proposed to strengthen control over the establishment of radio stations and frequency allocations. The problem was that there were now more stations than could freely operate on available frequencies and Hoover was attempting to find room for everyone by limiting station's power output and hours of operation.[8] Although bills were introduced in Congress embodying such restrictions, none were passed into law. The Secretary attempted to carry out the intent of the 1912 Act by inserting detailed conditions into the licenses, and declined renewals if conditions were not complied with. However, Hoover's attempts were seriously undermined when the United States Court of Appeals for the District of Columbia Circuit ruled that the Secretary of Commerce lacked legal authority for such actions, concluding that Congress had never intended to delegate such authority to the Secretary of Commerce thus leaving him powerless to deal with the situation.[9]

In July, 1926, as a stop-gap measure designed to prevent licensees from establishing property

---

[8]Gillmor and Barron, Mass Communication Law, p. 763.

[9]Erwin G. Krasnow and Lawrence D. Longley, The Politics of Broadcast Regulation (New York, New York: St. Martin's Press, Inc., 1973), p. 10.

rights in frequencies, both houses of Congress passed a joint resolution that no license should be granted for more than ninety days for a broadcast station or for more than two years for any other type of station. When Congress reconvened that December, the House and Senate quickly agreed on a comprehensive measure for the regulation of the radio industry. This Act, which became law in February, 1927, brought into existence the Federal Radio Commission.[10] At this point the telephone and telegraph industry had not yet been identified with the radio industry but was "regulated" separately, to a minor extent, by other elements of government.

a. Common Carriers Defined

Using the example of transportation, the Encyclopedia Britannica's discussion of carriers is subdivided into common carriers and contract carriers. Common carriers are defined as being those who "hold themselves out" to serve all; their charges, schedules, and routes are regulated, they are bound to serve all without discrimination and are entitled to a fair return on their investment; a "certificate of convenience and necessity" is required for operation and interstate business is subject to regulation by the

---

[10] Coase, "The Federal Communications Commission," p. 6.

Interstate Commerce Commission. Contract carriers differ in that they are not restricted to serving on fixed routes at regulated rates, except when the protection of the common carriers from such competition is essential to the public welfare.[11] At the time of the Radio Act and the FRC, the telephone and telegraph industries fit this definition of common carrier exactly.

The Federal regulation of business is based on Article I, section 8 of the Constitution, in which Congress is given the power "to regulate commerce. . . among the several states". Consequently, it is Congress that is primarily charged with the regulation of activities affecting interstate commerce.[12] This power has been delegated to "independent regulatory agencies" through general legislative statutes. Since communications by wire had grown up with the railroads, it had been placed under the regulatory jurisdiction of the Interstate Commerce Commission, not the FRC. However, during the period 1910 to 1934, the ICC had dealt with only eight telegraph rate cases, four

---

[11] Encyclopedia Britannica, 1973, Vol. 4, p. 965. Regulation protects the common carrier from his competitors but not necessarily from another technology.

[12] Gillmor and Barron, Mass Communication Law, p. 59. See the opinion of Mr. Chief Justice Vinson, American Communications Association CIO vs. Douds.

telephone rate cases and two cable rate cases.[13] With so little activity, it might be rightly said that actual government regulation of the telephone/ telegraph industry did not start until later. This can also be considered a bit tardy since, according to common carrier statistics, the assets of the American Telephone and Telegraph Company (AT&T) alone had reached more than $5 billion by 1934.[14]

b. Domestic Common Carrier History

In the early days of telephony through the 1880's, the Bell Telephone Company dominated the industry through a strong patent position, which it vigorously defended against all competitors. Seventeen years after telephone communications had originated there were 266,431 stations operating--all owned by Bell. The expiration of the basic telephone patents in 1893 and 1894 marked the end of the Bell System's complete monopoly over the telephone field and numerous independent telephone companies and manufacturers were formed. They offered competing services and stimulated the growth of the telephone industry. Less than

---

[13]Eston T. White, Utilities: Electricity, Gas, Telecommunications (Washington, D.C.: Industrial College of the Armed Forces, 1972), p. 113.

[14]Based on quoted common carrier statistics in Stuart L. Mathison and Philip M. Walker, Computers and Telecommunications: Issues in Public Policy (Englewood Cliffs, New Jersey: Prentice Hall, Inc., 1970), p. 2.

fifteen years later, the independent telephone companies owned 3.0 million stations compared to Bell's 3.1 million stations.[15]

However, in 1907, when Baker-Morgan banking interests gained control of the Bell system, Theodore Vail became its new president and reversed a number of Bell policies, emphasizing absorption of the competition.[16] Now called American Telephone and Telegraph (AT&T), the company had accumulated enough local operations to take over the industry simply by wielding financial and political power.[17] AT&T initially divided the industry with Western Union, telephone for the former and telegraph for the latter. Having thus neutralized its strongest telephone competitor by this action, it consolidated long distance networks and began to absorb the independents who were unable to compete. AT&T soon dominated the long distance service and no regulations or genuine authority existed at that time which required them to provide for interconnection with independent systems that remained.

---

[15] Richard Gable, "The Early Competitive Era in Telephone Communication, 1893-1920," Law and Contemporary Problems, 34 (Durham, North Carolina: Duke University, 1969), pp. 343-344.

[16] Ibid., p. 345.

[17] "Independent Phone Companies: The Best Kept Growth Secret," Business Week, May, 1973, pp. 86-87.

The Bell system's acquisition attempts were strongly resisted by the independents; but only through threatened nationalization did the government, during the Wilson Administration, stop AT&T's rout of the independents. AT&T, in varying degrees, had refused to interconnect with independent exchanges for long distance service. The independents, complaining to Attorney General George Wickersham, charged Bell with antitrust violations. The complaints were resolved by the Kingsbury Commitment of 1913, which was an AT&T-offered compromise that in reality had no impact on its dominant position in the industry.[18]

2. The Federal Communications Commission

In response to a request from President Roosevelt for a study of the organization of radio regulation, in January 1934 Secretary of Commerce Daniel Roper issued a report recommending the consolidation of the communications regulatory activities of the FRC, the Interstate Commerce Commission (ICC), the Postmaster-General, and the President into "a new or single regulatory body

---

[18] Gable, "The Early Competitive Era," pp. 352-353. Under this commitment, which was an agreement between the Attorney General and AT&T Vice President N. C. Kingsbury (drafted by Kingsbury), the Bell System agreed not to acquire control over any competing company (but it did not restrict Bell from acquiring noncompeting telephone companies!) and to interconnect independents with its own system providing the former's equipment met Bell standards (between 1913-1917, Bell purchased over 241,000 stations from independents and sold 58,000 stations).

to which would be committed any further control of two-way communications and broadcasting."[19] The groundwork was thus laid for Congressional action and the Communications Act of 1934 was passed.[20]

The Federal Communications Commission is the creature Congress created by that Act to execute and enforce its provisions.[21] Originally intended to regulate the fledgling radio industry,[22] the Act also made various organizational changes to the Federal Radio Commission and gave the agency broad powers over all communications, including telephone and telegraph (Title III of the 1934 Act, which dealt with radio, was almost identical with the Radio Act of 1927).[23] The language was broad in scope and was capable of

---

[19]Krasnow and Longley, The Politics of Broadcast Regulation, p. 14. For further details see Senate Committee Print, S. Doc. 144, Study of Communications by an Interdepartmental Committee, 73rd Congress, 2nd session, 1934.

[20]The Communications Act of 1934, with Amendments and Index Thereto (Washington, D.C.: U.S. Government Printing Office).

[21]Krasnow and Longley, The Politics of Broadcast Regulation, p. 7. Also see Title I, section 1 of the Act.

[22]Bary Taub, "Federal Communications Commission Regulation of Domestic Computer Communications: A Competitive Reformation," Buffalo Law Review, 22 (Spring, 1973), p. 951.

[23]Krasnow and Longley, The Politics of Broadcast Regulation, pp. 11-14. The Act defined a common carrier as "any person engaged as a common carrier for hire, in interstate or foreign communication by wire or radio."

application to a host of other activities.[24] The Act also established that the Commission's powers were not limited to the engineering and technical aspects of regulation of radio communications but rather to the "larger and more effective use of radio in the public interest."[25]

Congress acted upon the knowledge that if the potentialities of radio were not to be wasted, regulation was essential. The facilities of radio were not large enough to accommodate all who wished to use them. Methods were needed for choosing from among the many who applied. Congress itself committed this task to the Commission providing as a touchstone the "public interest, convenience or necessity."[26]

a. The Public Interest

As far as domestic common carrier regulation is concerned, the "public interest" factor seems to be

---

[24]Taub, "FCC Regulation of Domestic Computer Communications," p. 951. The applicability of the Act to communication satellites was addressed at the time of COMSAT and again during the DOMSAT proceedings.

[25]Gillmor and Barron, Mass Communication Law, pp. 764-765.

[26]Coase, "The Federal Communications Commission," pp. 12-13. Extracted from Mr. Justice Frankfurter's opinion in NBC v. United States, 319 US 190, 213 (1943). The concept of "public interest" was introduced into American legal thought by Munn v. Illinois, 94 US 113 (1876) and is further discussed in Loevinger, "Regulation and Competition as Alternatives", Chapter III, note 154, pp. 128-135.

something recognizable but difficult to define. Former FCC Chairman Dean Burch, in a speech before the American Bar Association, defined the public interest as those actions which:

> create a prevailing climate in which the widest possible range and variety of services are provided to the public by the greatest practical number of independent entities, each one seeking to satisfy public wants in its own way.[27]

By this definition, "public interest regulation" appears to be less than twenty years old, even though the regulator and his charter have existed for over twice that long and the telephone and telegraph industries have existed for over five times that long. Even the brief history of the industry's development prior to 1934 that has been presented shows that government regulation merely gave official approval to the historical accidents that had shaped the business and failed to provide national guidelines. A cursory look at the development of the industry from 1934 to 1959 lends additional support to this view. There was in fact little demonstrated action "in the public interest" shown by the carrier regulators prior to 1959, when the Commission's policy of increased

---

[27] Dean Burch, "Public Utility Regulation: In Pursuit of the Public Interest," Public Utilities Fortnightly (September, 1973), p. 70. Burch was Chairman during the key years for DOMSAT policy formulation (October 31, 1969 to March 8, 1974).

competition was adopted.[28]

The public interest considerations in the use of communications satellites involve more than just the question of trying to develop competition in the interest of the consumers. Some believe that it is important to secure maximum utilization of satellite systems to accomplish purposes in education and health, and other fields which economically are unprofitable but which have great social implications.[29] Others see the satellite as a means to break AT&T's monopoly of the common carrier industry and as a cost-cutting alternative to existing long-distance costs. The "public interest" in satellites means many things and the diversity of congressional opinions on the subject of satellite communications, which is discussed in Chapter II, provides an excellent example of this. Since the issues surrounding DOMSAT were as complex as the common carrier industry itself, a brief description of the Nation's primary domestic communications carriers is believed to be necessary

---

[28]This refers specifically to the Commission's Above 890 decision, when private ownership of microwave communications systems was first authorized. For more details see notes 74 and 76, *infra*. This seems to be an agreed upon milestone; *see* Taub; note 22, *supra*, p. 963; also Trebing, note 40, *infra*, p. 309.

[29]Thomas P. Murphy, "Technology and Political Change: The Public Interest Impact of COMSAT," *The Review of Politics*, *33* (July, 1971), p. 424.

for a better understanding of the DOMSAT discussions.

b. The Regulated Common Carriers

The magnitude of today's telephone and telegraph systems is something that may not be visible to the average user. People often refer to Bell Telephone or Western Union as big and think no more of it. One contemporary viewpoint sums up competition, regulation, and the nation's telephone industry as follows:

> First of all, capitalism is the best. It's free enterprise, right? Barter. . . Communism is like one big phone company; government control, man. And if I get too rank with that phone company, where can I go, man? I'll end up like a schmuck with a Dixie cup and a thread.[30]

An uninformed public can quickly relate to such commentary and for good reason. A current magazine advertisement reads:

> The Bell System. It's an incredible operation. It takes a mind-bending multitude of cables and switches and gear to make all 114 million telephones talk to each other. It takes a master plan to keep this system running 24 hours a day. It takes a totally unified system to make it all work together. . . . The result of all this planning is, quite simply,

[30] Monologue by Lenny Bruce at his famous Midnight Concert, Carnegie Hall, February 4, 1961. Bruce was a comedian of the 1960's who addressed controversial situations found within our society that he felt should be spoken about, reflected upon and improved upon. While commenting on Capitalism versus Communism, Bruce found that the telephone monopoly served as a descriptive example for him and as a problem/situation in its own right worthy of his comment.

the best phone system in the world.
One Bell System. It works.[31]

The facts reveal that there are 1,785 landline telephone companies in the U.S. with operating revenues totaling more than $25 billion, with plant assets in excess of $84 billion, and approximately one million employees. Sixty-one of these carriers provide comprehensive reports to the Commission.[32] At the time of the Kingsbury Commitment, AT&T had been servicing about 5.1 million telephones while some 20,000 independent telephone companies were serving about 3.6 million telephones.[33] Today, in comparison, AT&T services approximately 109 million telephones while the 1500 independents serve the remaining 24 million telephones of the nation's system.[34] A rough breakdown of the industry is as follows:

(1) AT&T (23 operating companies)--82 percent

(2) GT&E (30 operating companies)--8 percent

(3) Eleven holding companies and large

---

[31]"Bell System Advertisement," Smithsonian, Vol. 6, No. 7 (October, 1975), pp. 31-32. As this is a national magazine, the reader might easily misinterpret the "totally unified system" as being his or her national telephone system.

[32]U.S. Superintendent of Documents, 39th Annual Report/Fiscal Year 1973, Federal Communications Commission (Washington, D.C.: Government Printing Office, Stock No. 0400-00284), p. 287.

[33]Borchardt, Structure and Performance of the U.S. Communications Industry, p. 29.

[34]"Independent Phone Companies: The Best Kept Growth Secret," p. 84.

independents--6 percent

(4) 1,500 small independents--4 percent

The Bell companies serve approximately one-third of the geographical area of the United States and the independents serve a second third of the country. The remaining third is too sparsely populated to economically service by the traditional means of wire and cable.[35]

The Western Union Telegraph Company is basically the sole domestic telegraph carrier. Dollarwise, it is less than two percent the size of telephone system but because of the fact that it provides a specialized service (in the form of record communication and custom-built private systems), it is actually the forerunner of the specialized carrier industry that has been developing during the past five years and second only to AT&T in national importance.[36] Western Union is also the proud owner of "Westar", the first domestic satellite system which was put into orbit on

---

[35]"Independent Phone Companies: The Best Kept Growth Secret," note 17, <u>supra</u>, p. 85. The independents have 11,000 central exchanges while the Bell System companies have just over 12,000. The Bell exchanges are predominantly located in large metropolitan centers, each serving an average of 21,800 phones. The independent exchanges serve an average of only 1,570 phones. See Mathison and Walker, note 14, <u>supra</u>, p. 113 for further details.

[36]Mathison and Walker, <u>Computers and Telecommunications</u>, p. 6. The specialized carriers are discussed in Chapter III since they evolved during the time that DOMSAT policy was being formulated.

April 13, 1974.

The Bell System owns approximately 98 percent of the Nation's long-distance facilities, which interconnect the individual telephone companies together across state lines.[37] Such business is considered interstate commerce and falls under the jurisdiction of the FCC. AT&T Long Lines, the responsible Bell operating company in this area, has not been subjected to antitrust laws because of its holdings but has instead been shielded by the protection of regulation. Long Lines actively recruited this regulatory shelter for many years and its monopoly status actually predates the onset of regulation.[38] Because of their insensitivity to distance, satellites have quickly become economically competitive in this area and have threatened to modify the industry's structure. In self defense, the established carriers have assumed a variety of positions designed to neutralize and minimize the effects of communications satellites on established markets.

Although the satellite in space represents

---

[37]Harvey Averch and Leland L. Johnson, "Behavior of the Firm Under Regulatory Constraint," American Economic Law Review, 52 (1962), p. 1060.

[38]William Melody, "Technological Determinism and Monopoly Power in Communications," Presentation before the American Economic Association, New Orleans, December 28, 1971. Other supporting historical material can be found in Gable, note 15, *supra*.

probably the most novel means of communications yet devised by man, the domestic common carriers initially looked upon this innovation as no more than a "telephone pole in the sky". However, when this technique was implemented for international communications, the potential impact on the domestic market became evident and the Commission's responsibilities were increased and expanded, as the international lawyers are fond of saying, "ad caelam"--to heaven itself.[39]

B. The Constraints of Business and Technology

The market structure provided a setting in which the existing common carriers, on the one hand, and the potential entrants on the other, pursued conflicting courses of action, subject to the constraints of highly imperfect markets and dynamic technologies.[40] But any policy must operate within existing technical constraints; policy making only begins at this stage. Revolutionary shifts in technology and aggressive innovation may be aborted if they do not receive the support of thoughtful public policy.

Organizational forms which would permit the

---

[39] Rosel H. Hyde, "Space Age Regulation," Presentation before the Midwest Association of Railroad and Utilities Commissioners (Chicago, Illinois: July 7, 1964), p. 3.

[40] Harry M. Trebing, "Common Carrier Regulation--the Silent Crisis," Law and Contemporary Problems, 34 (Durham, North Carolina: Duke University, Spring, 1969), p. 318.

greatest development of the technology and the widest play of operating alternatives had to be considered. This was critical since policy decisions that impact on the market structure (and the respective roles of competition and regulation) once made, are not easily reversed.[41] To a large degree many of the problems that faced the domestic industry were associated with the pressures for change that arose from the technological advance and the economic growth of the postwar years.

1. The Market Structure

The Bell Telephone System, the independent or non-Bell telephone companies and Western Union operate virtually all of the nation's common carrier telephone and telegraph facilities. The telecommunications industry had developed under conditions of the so-called natural monopoly. Entry of new suppliers was restricted, if not foreclosed, with the result that competition was almost absent as a market force. The regulatory agency, for the most part, had confined itself to a concern for the economic well-being of the regulated industry and to the correction of excesses in pricing practices.[42] But what could be considered

[41]Gabel, "The Early Competitive Era," p. 359.

[42]Bernard Strassburg, "New Technology and Old Institutions," Telecommunications (June, 1974), p. 23. Strassburg, a thirty-year veteran of the FCC, was the Common Carrier Bureau Chief for the period 1964-1974.

a natural monopoly in some static efficiency sense might also be considered an "unnatural" one in terms of meeting the prerequisites for innovation and growth.[43]

Historically, it had been assumed that communications services were provided under conditions of natural monopoly, although the basis for this has never been made explicit.[44] Since World War II the consolidated voice communications market had shown remarkable stability, increasing at an average annual rate of eight percent, the greatest imponderables were the demands for new services such as data and video transmission.[45]

The stakes were high for everyone involved since the horizontal market (that is, the percentage of all households and business firms with telephones) was rapidly approaching saturation.[46] If the Bell

---

[43]Alfred E. Kahn, The Economics of Regulation: Principles and Institutions, Vol. I (New York: John Wiley and Sons, 1971), p. 12.

[44]Manley R. Irwin and Harry M. Trebing, "A Survey of Problems Confronting the Communications Industry in the United States," Telecommunications for Canada: An Interface of Business and Government (Toronto, Canada: Methuen Publishing, 1973), pp. 214-217.

[45]Ibid., p. 216

[46]Department of Commerce statistics show that 92 percent of all homes in the United States have telephone service today. See U.S., Superintendent of Documents, Statistical Abstracts of the United States, 95th Annual Edition, 1974 (Washington, D.C.: Government Printing Office, stock no. 9324-00108, 1974).

System and the common carriers failed to establish a strong foothold in these future markets, they could look forward to drastically reduced rates of growth and a significant shrinkage of their relative importance.[47]

a. Market Economics

Economics deals with the allocation of limited resources towards satisfaction of unlimited wants. Resources are typically identified as land, labor and capital plus a technology that determines their transformation into consumer goods.[48] The technology is viewed as a parameter like the weather, affecting the outcome of resource allocations but itself unaffected by them.[49]

The domestic telecommunications industry is characterized by rapid technological advance interacting with market changes in the level and composition of demand.[50] However, it has also been demonstrated that

---

[47]Trebing, "Common Carrier Regulation," pp. 310-311.

[48]The fact that a public interest balance may not be achieved via the market system in the presence of monopoly elements provides an economic rationale for antitrust laws.

[49]Jacob Schmookler, "Technological Change and the Law of Industrial Growth," Patents and Progress; the Sources and Impact of Advancing Technology (Homewood, Illinois: Irwin Publishing, 1965).

[50]Leland L. Johnson, "Technological Advance and Market Structure in Domestic Telecommunications," American Economic Association, 60 (May, 1970), p. 204.

the quest for profit is also a primary influence on the rate and direction of innovation, despite the large role of other goals motivating discovery, that must be considered. Moreover, the relationship appears bidirectional, with the state of knowledge shaping and being shaped by profit opportunities and availability of resources.[51] It is certain that the prospect of being permitted to enter an established multi-billion dollar industry for the purpose of competing with the established monopoly of that industry by means of a new technology stirred many a corporate heart.

Utility sectors commonly proceed through four stages, as elasticities of demand vary. In stage one, the system is invented, often leading to control by patents. It is usually a brief period but decisive for the form of the system. Stage two involves the system's creation and growth; often the system is displacing a prior "utility". Cross-subsidies are involved and the service usually seeks regulated status for permanence, legitimacy and market control. In stage three, the system becomes complete as a function of technology and market saturation and it shifts from the offense to the defense, competing with new technologies and challenged by the users. Finally, in

---

[51]Morton I. Kamien and Nancy L. Schwartz, "Market Structure and Innovation: A Survey," Journal of Economic Literature, 13 (March, 1975), p. 31.

stage four, the system yields to the pressures of competition and technology and, now no longer a utility, reverts to conventional competitive procedures.[52]

William Shepherd believes that the telephone industry has been in stage three since 1947. Was it not possible that DOMSAT could be the means that would potentially break the back of the AT&T monopoly? Dr. Burton A. Kolb, a Professor of Finance at the University of Colorado, has noted that:

> A public utility usually faces severe competition only twice in its life, once when it rises to prominence and again when it is superceded by a superior technology. In contrast the industrial enterprise is subject to the continual interaction of competitive forces, including technological change. But these forces rarely are of such magnitude as the technological revolution which seriously impairs or destroys the economic value of the public utility.[53]

Domestic satellites posed such a threat to the common carrier market. Satellite technology possessed a glamour that attracted widespread public interest as well as the potential for new, better and cheaper communications services. Communications satellites threatened to change the traditional role of the

---

[52]William G. Shepherd, "Entry and Communications," _Competition and Monopoly in the Domestic Telecommunications Industry_ (Lexington, Virginia: Washington and Lee University, 1974), pp. 38-39, 59.

[53]Burton A. Kolb, "The Rise and Fall of Public Utilities--An Appraisal of Risk," _The Journal of Business_, _37_ (1964), p. 343.

domestic carriers.[54] New markets and new potential suppliers raised the possibility of rendering obsolete the traditional concept of "natural monopoly", a phrase that Professor James R. Nelson of Amherst labeled as "one of the most unfortunate. . .ever introduced into law or economics. . .", believing that "every monopoly is a product of public policy."[55] Looking at the regulatory trend of the 1960's, the Commission was definitely working toward increased competition [the Interconnect (1968) and Specialized Carrier (1971) decisions are discussed in Chapter III]. While there was some apprehension that under certain conditions the common carrier would have an incentive to operate at a loss in competitive markets and shift financial burden to its other services,[56] others felt that the regulatory agency should take advantage of whatever

---

[54] Johnson, "Technological Advance and Market Structure in Domestic Telecommunications," p. 204.

[55] Charles F. Phillips, Jr., "Domestic Telecommunications Policy: An Overview," Washington and Lee Law Review, 29 (1972), pp. 235-236. The emergence of satellite transmission in the 1960's has been paralleled by advances in land-based alternatives for high-density routes. Discussed further in Specialized Carriers, Chapter III.

[56] Averch and Johnson, "Behavior of the Firm Under Regulatory Constraint," p. 1065. In a review of the commentary on the A-J effect model, Johnson has shown that the earlier analysis remains theoretically valid. Also noted is that the FCC is more inclined to open the telecommunication field to competitive forces [than in 1962] and is no longer willing to accept uncritically the argument that a natural monopoly exists in all cases. See L. Johnson, "The Averch-Johnson Hypothesis

competitive possibilities existed.[57] New trends in demand and technology suggested that several parts of the point-to-point [as opposed to broadcast] communications industry might be amenable to even a fully competitive structure, particularly the large-scale transmission of data and domestic satellites as an alternative to land-based transmission.[58]

b. Rate of Return Regulation

Rate of return regulation, in conjunction with the market structure, can give rise to distorted investment decisions. In establishing the level of prices charged by public utilities, regulatory agencies commonly employ a "fair rate of return" criterion, which is computed as the ratio of net revenue to the value of plant and equipment (the rate base).[59]

After Ten Years," Regulation in Further Perspective (Cambridge, Massachusetts: Ballinger Publishing Company, 1974), pp. 75-78.

[57]Donald F. Turner, "The Scope of Antitrust and Other Economic Regulatory Policies," Harvard Law Review, 82 (1969), p. 1207.

[58]William G. Shepherd, "The Competitive Margin in Communications," Technological Changes in Regulated Industries (Washington, D.C.: The Brookings Institute, 1971), pp. 86-89. It is the combination of these two parts that proves particularly interesting. It is this combined market to which the SBS filing has addressed itself (see Chapter IV). For an economic analysis of the early DOMSAT proposals using published economic models, see Stuart N. Goodman, "An Analysis of Domestic Satellite Communications in the United States," (thesis, Polytechnic Institute of Brooklyn, June 1968), 239 p.

[59]See Averch and Johnson, "Behavior of the Firm Under Regulatory Constraint," pp. 1052-1069 for an extensive discussion of the economic considerations.

Therefore what goes into making up the rate base is very important to the carrier. His incentives as a monopolist may be to retard the use of his inventions in favor of more costly technology, to engage in more inventive activity than an equivalent unregulated carrier, or to allow excessive requirements of reliability and quality to shape the whole direction of his technology.[60] Because regulation limits his rate of return, he may tend to choose a more capital intensive technology and enlarge his rate base.

The mere fact that a new entrant's rates for a particular route or a particular service are lower than those of the established carrier does not indicate that the new entrant's costs are necessarily lower than the existing carrier's long-run incremental costs for comparable service. In order to discourage uneconomical entry, it is essential to permit the carriers to respond by adjusting their rates toward their own incremental costs. Existing rates must not be frozen to provide an umbrella protecting uneconomical competitive activity. However, at the same time the danger

---

[60]Shepherd, "The Competitive Margin in Communications," pp. 88, 96-97. High reliability to avoid embarrassing service outages may extend the carrier's technology far beyond the optimum. N. E. Feldman of Rand is presently looking into what are realistic criteria for reliability vs. requirements in communications systems. Personal interview with Mr. Nathaniel E. Feldman, Research Engineer, The Rand Corporation, May 4, 1976.

exists of a carrier cutting prices to the point where revenues fall even below incremental cost in particular competitive markets if it has protected revenues from other markets.[61]

Therefore, the carriers may have special incentives to "select" innovations, to invoke regulatory procedures, and to control the flow of technological information so as to minimize the probability of new entry into any of their actual or desired markets.[62] They have been seen in the past as slow to innovate and introduce new techniques and facilities.[63] In the case of AT&T, its high inertia is particularly bad in many respects. Especially during the last decade, the legal monopoly has bitterly resisted many innovations that later proved beneficial to the users in general and neutral or even beneficial to AT&T itself.[64]

---

[61] Johnson, "Technological Advance and Market Structure in Domestic Telecommunications," pp. 205-206.

[62] Shepherd, "The Competitive Margin," p. 86.

[63] Mathison and Walker, Computers and Telecommunications, p. 146.

[64] "The Five Dollar Phone and Other Fears," Modern Data (February, 1975), p. 28. Sebastian Lasher, technical consultant to FCC Commissioner Washburn and formerly with the OTP, attributes such resistance to the Maslow School of Psychology, "if the only tool you have is a hammer, you tend to treat everything as if it were a nail." See "Remarks of Sebastian A. Lasher, Office of Telecommunications Policy," International Communications Association Conference, May 10, 1974, p. 7.

Depreciation policies are another example of the type of decisions that can contribute to an inflated rate base. Depreciation should reflect the economic cost of providing service and should include an allowance for obsolescence caused by technological advance. The depreciation policies of AT&T are based on the straight line methodology, the use of which does not appear to reflect the economic realities of a dynamic industry undergoing rapid technological change.[65] In establishing a rate base there can be hundreds of accounting decisions that the carrier will make that will affect his rate of return and the cost to the customer. As former Commissioner Nicholas Johnson noted:

> In an industry whose annual revenues are roughly twice the yearly income tax collected by all fifty states combined, a fraction of a percent here and there may amount to millions of dollars in phone bill savings.[66]

Convincing arguments exist which show that conservative straight-line depreciation for ratemaking purposes will maximize the rate base and minimize the current charge to expenses.[67] This may result in

[65] Kenneth B. Stanley, "International Telecommunications Industry: Interdependence of Market Structure and Performance Under Regulation," Land Economics (November, 1973), pp. 398-399.

[66] Nicholas Johnson, "Why Ma Bell Still Believes in Santa," Saturday Review, March 11, 1972, p. 60.

[67] Irwin and Trebing, "A Survey of Problems," p. 228.

politically popular service rates, but it may also constitute a major barrier to innovation and technological advance.

A 1972 Business Week article summarized these arguments in a critique of depreciation policies:

> [I]n figuring depreciation, Bell takes very long equipment lifetimes. For example, New York Telephone writes off the cost of an electronic central office over 38 years, so it gets its investment back at the almost invisible rate of 2.6% a year. As an over-all average, AT&T depreciates its plant at a little more than 5% a year.
>
> From an accounting standpoint, the computer industry, which is also capital-intensive and service oriented, looks altogether different. Almost all computer makers capitalize only the manufacturing cost of the equipment they put out on rental, or about 20% of what they would get for it in an outright sale. They write off installation and customer service costs immediately as expenses.[68]

Such rapid write-offs encourage the use of new technology and represent the opposite extreme of the common carriers' accounting practices, which discourage the retirement of obsolete equipment and hence discourage the application of new technologies.[69]

c. Competition and Antitrust

Competing technology and the growth of new services posed several issues which challenged the

---

[68] Business Week, March 25, 1972, pp. 57-58.

[69] Ibid.

assumptions of market structure long associated with the communications industry. These forces confronted the regulator with two policy alternatives. The first policy choice was to protect existing competitors, or more specifically to opt for a market status quo. The second choice was to employ market structure as a means to exploit either new technical developments, new communications markets, or both. History and the FCC's activities since 1959 clearly show that the second choice was the chosen policy. Nevertheless, concern for antitrust was always in evidence also. This is important since fear of antitrust involvement can act as a constraint to major companies, like IBM, on any plans which such companies might consider in the field of communications, leaving the planning of new services or alternative methods for existing services either to the existing carriers or to companies which have fewer commitments.[70]

(1) Competition. In the telecommunications industry, competition has been a consideration since the early days of telegraph, when international overseas communications services were provided by

---

[70]Borchardt, Structure and Performance of the U.S. Communications Industry, p. 141. Borchardt raised this thought in 1970. Perhaps the fear of antitrust complications was a significant factor in keeping IBM from seeking the opportunity to apply its technological expertise to the domestic communications area until 1974.

undersea cable. In 1927 high frequency radio made possible for the first time both overseas telegraph and telephone service.[71] When this technology was first applied by companies interested in its commercial exploitation, Congress was persuaded that this new technology should be permitted to compete effectively with the older telegraph cable technology. Consequently the Radio Act of 1927 prohibited mergers of carriers-by-radio with carriers-by-cable if the purpose or effect of such mergers was substantially to lessen competition. This prohibition was designed to protect the development of the new technology which required less capital from being slowed down by the older cable technology which required larger capital investments. This was reenacted as section 314 of the Communications Act of 1934.[72]

After the end of World War II, the demand for new types of bulk communications services, combined with advances in microwave radio technology, confronted policy makers with a variety of issues challenging the structure of the telecommunications industry.[73] In 1959 the FCC's _Above 890_ decision removed all

---

[71]Undersea cables for telephone did not come until almost thirty years later.

[72]Borchardt, _Structure and Performance of the U.S. Communications Industry_, pp. 46-47.

[73]Ibid., p. 25-26.

significant barriers to the installation and operation of private microwave systems. The Commission found no basis for concluding that the licensing of private communications systems would adversely effect the ability of common carriers to provide service to the general public or that it would adversely effect the users of such common carrier services.[74] Although carriers could offer the communications service at a lower rate than private firms because of the economics of scale and the shared use of facilities, the Commission felt that the opportunity to introduce "competition" in the nation's system outweighed the small social loss due to diseconomics of scale and the nominal adverse effects upon carrier revenues.[75] The

---

[74]Phillips, "Domestic Telecommunications Policy," pp. 239-240. In 1956 prospective private users asked the FCC for access to radio frequencies above 890 megahertz to develop noncommon carrier microwave service. The suppliers of microwave equipment joined in the request. The existing policy at the time licensed private microwave communication systems to government and business units only when they had "special communications needs," such as a lack of common carrier facilities.

[75]Mathison and Walker, Computers and Telecommunications, p. 117. In this case, competition is referring to the supplying of communications equipment, not services. The 1956 Consent Decree of AT&T had previously eliminated two possibilities of new competition in this area. No way was provided for outside suppliers to compete in the sale of equipment to Bell operating companies and at the same time, Western Electric agreed to produce only telephone equipment for Bell system and government use. Since then it has not bothered to enter foreign markets, regarding itself as "fully occupied" with domestic responsibilities. See note 83, *infra*.

seeds of competition were planted.[76]

Competition, or more properly economic competition, implies more than just the vying for customers or markets. It also means the absence of monopoly, on either the buying or the selling side, and the absence of government intervention in the market process. It denotes a sufficient number of well informed, independent competitors so that no individual can affect the market by restricting sales or purchases. Relatively easy entry into or exit from the market must also be possible.[77] The obvious trend in FCC policy since 1959 has been towards "competition" in one way or another. But because entry into the carrier industry is determined by the Commission only,[78] the established carriers choose to call it a policy of "regulated competition," giving it negative connotations. In any event "competition" was the Commission's policy throughout the satellite issue and remains that today.

---

[76]The case was hailed as a landmark decision that "may well determine the depth of competition" in the communications industry for several decades to come. See Taub, note 22, _supra_, p. 963.

[77]_Encyclopedia Britannica_, Vol. 6 (1973), p. 234.

[78]See the FCC's 39th Annual Report/Fiscal Year 1973, note 32, p. 108. Commission approval is required before a carrier may construct, acquire or operate facilities and before it can discontinue or curtail services. Mergers, consolidations and acquisitions of property of one carrier by another must also be passed on by the Commission.

The term "competition" has aroused more emotion in connection with common carrier matters before the FCC than any other word or phrase in recent memory. Depending on one's frame of reference, it is considered either disastrous, disruptive or terrific for the communications industry. As long as the pros and cons of competition were being argued in FCC hearing rooms and Federal court rooms, none of the contentions advanced could be proven or disproven. However, since the Commission adopted policies fostering competition the action has shifted to the marketplace.[79]

(2) Antitrust. In addition to direct regulation by the Commission and its predecessors, the domestic communications industry has been the subject of antitrust action on a selective basis more than 60 years.

The first antitrust suit was threatened by the Justice Department in 1913. As noted earlier, the independent telephone companies charged that Bell refused to provide satisfactory long-distance interconnections. In response to this pressure, Bell entered into the Kingsbury Commitment which set forth minimum concessions only.[80]

---

[79]Kenneth W. Gross, "Competition is Not a Dirty Word," Telecommunications, Vol. 10, No. 4 (April, 1976), p. 39.

[80]See note 18 *supra*.

In 1921, the Willis-Graham Act permitted telephone companies to merge or consolidate with competing companies subject to approval by state commissions and the ICC. This Act effectively terminated the Kingsbury Commitment and Bell again embarked on a program of acquisition. These efforts led to complaints by USITA (United States Independent Telephone Association). As a result, AT&T Vice-President E. K. Hall set forth Bell's policy on horizontal mergers in a memorandum in 1922 to the President of USITA. The Hall Memorandum stated that Bell was opposed to further acquisitions of the independents as a general policy, except in "special cases", which were broadly defined in terms of public convenience and service.[81]

AT&T is an excellent example of a holding company. It exercises control through stock ownership over some 23 operating or associated companies throughout the United States; it owns 100 percent of the stock of Western Electric, which accounts for some 85 percent of the domestic communications equipment market, and shares ownership with Western Electric of the Bell Laboratories, the research arm of the company.[82]

A major assault on the vertical relationships

---

[81]Irwin and Trebing, "A Survey of Problems," pp. 222-223.

[82]Ibid., p. 215.

of AT&T and Western Electric occurred in 1949. In that year the Justice Department filed a suit alleging that Western Electric had, in monopolizing the manufacture and supply of communications equipment and apparatus, violated Section 2 of the Sherman Antitrust Act. The Government sought as its remedy both the divestiture of the Bell-Western Electric relationship and dissolution of Western Electric into three competing firms. It was hoped that this would introduce competition in the manufacturing and supply of related communications equipment. The suit ended in a 1956 consent judgment where AT&T was required to make its patent portfolio available on a royalty-free basis and technical information available to outside suppliers. However, the decree, in sanctioning the existing AT&T-Western Electric structure, preserved the vertical relationship of telephone carrier and telephone manufacturer.[83]

2. Satellite Technology

Artificial satellite technology, which established one of two major technical boundaries for DOMSAT, is less than twenty years old. The Soviet Union announced on October 4, 1957 that it had successfully launched the first manmade satellite into orbit around the earth. Sputnik I, as it was called, reportedly carried

[83]Ibid., p. 223.

184 pounds of scientific instruments and circled the earth every 96.2 minutes.[84] The first step necessary for exploiting Arthur Clark's idea of communications relayed by satellite had been taken. But by 1961, it still seemed doubtful whether rocketry would achieve such accurate positioning in the near future or whether small solar-powered electronic devices could be used to establish noise-free communication links as Clark had perceived it.[85]

Clark had envisioned a system which would use three satellites, orbiting the earth in geostationary orbit, and could relay point-to-point or broadcast communications to any location on the globe. The geostationary orbit is the band of space in which satellites circle the earth at a speed equal to its rotation and appear to hang motionless above a fixed point on the Earth's surface. This band lies 22,300 miles above the equator and the number of satellites which can be accommodated along this orbit is a major determinant of potential satellite communications capacity. From its apparently stationary position

---

[84]Lloyd D. Musolf, Communications Satellites in Political Orbit (San Francisco: Chandler Publishing Co., 1968), p. 14.

[85]James Martin, Future Developments in Telecommunications (Englewood Cliffs, New Jersey: Prentice-Hall, Inc., 1971), p. 222. Orbit locations in outer space were regarded as a hostile environment for electronic circuitry. Today they seem more benign.

above the surface of the Earth, a synchronous satellite has approximately forty percent of the surface of the Earth constantly in view and can provide line-of-sight communications between any two points on that surface. The satellite has in this way introduced a new dimension into communications technology.[86]

Frequency spectrum utilization established the second major boundary condition. The portion of the spectrum which is used for radio transmission is actually very small. Although the radio spectrum is considered to range from ten kilohertz (10,000 cycles per second) to three terrahertz (3 million million cycles per second), only 40 gigahertz (40,000 million cycles per second) had been allocated through international agreement in the 1960's.[87] This is equivalent to less than 7,000 television circuits. Although frequencies as high as 300 gigahertz (GHz) are sometimes used for experimental purposes, physical existence of the spectrum does not mean that it is technologically or economically useable. The higher the frequency the more sophisticated the technology used must be. Propagation characteristics of radio waves vary with frequency also and with

---

[86]Roscoe L. Barrow and Daniel J. Manelli, "Communications Technology--A Forecast of Change (Part I)," Law and Contemporary Problems, 34 (Durham, North Carolina: Duke University, 1969), p. 216

[87]Ibid., p. 208. A discussion of the International Telecommunication Union and the internation frequency allocation process is found in Chapter III.

satellites in space, the attenuation and scattering of signals passing through the atmosphere and the ionosphere must be taken into account.

a. Orbit Considerations

The orbit of a communications satellite affects the service that can be provided since it determines the amount of time a satellite will be visible to an Earth station, in what locations these Earth stations will be and how complex and expensive they must be.

Clark's geostationary approach is one option. However, even if placed at the correct height and having the right velocity, the satellite will not remain stationary because of the Earth's equatorial ellipticity and perturbations resulting from movements of the Sun and the Moon. Corrections to height and velocity are required at regular intervals throughout the life of the satellite and, since it is not economic to correct the satellite too frequently, system design must allow for drift over a period of months.[88]

Random orbits, polar orbits and inclined elliptical orbits, using low to medium altitude satellites, are examples of nonstationary techniques. Although systems using these techniques each have applications where they provide specific advantages, such systems

---

[88]B. J. Halliwell (ed.), Advanced Communication Systems (London: Butterworth, 1974), pp. 208-209.

normally require multiple satellites for continuous coverage and expensive Earth station tracking systems. The average costs of a sophisticated Earth station today has been estimated at approximately $4.5 million. During the early considerations of satellite communication system alternatives, Bell had supported random-orbital technology although the distinctly less capital-intensive synchronous-orbit method was a possible alternative. Synchronous systems were adopted as preferred quickly after 1963 primarily because the technology necessary for deploying such systems had been perfected by an outsider, Dr. Harold Rosen of Hughes Aircraft, and was being promoted for competitive reasons.[89]

The coverage of a synchronous satellite varies with its location in the geostationary orbit and the restrictions on minimum elevation angles at the Earth stations. The minimum elevation restrictions arise from signal quality factors and increased coordination problems with terrestrial systems at the lower angles. For elevation angles greater than five degrees the

---

[89] Shepherd, "The Competitive Margin in Communications," pp. 105-106. By some estimates, this occurred at least five years and hundreds of millions of dollars sooner than the carriers would otherwise have achieved it. See Lawrence Lessing, "Cinderella in the Sky," <u>Fortune</u>, <u>76</u> (October, 1967), pp. 198, 201. Shepherd also notes that unfortunately there are no detailed studies of these savings in the public record. See also Kahn, Vol. II, <u>The Economics of Regulation</u>, p. 67.

useful arc for coverage of the contiguous U.S. ranges from about 53°W to 138°W longitude or approximately 85°. If the elevation angle restriction is increased to 10° the useful arc is reduced to approximately 70°.[90] The separation of satellites on 70° of geostationary arc would be no problem if each satellite could use different portions of the frequency spectrum. But the problem is not one of physical space but one of available spectrum and of frequency interference.

b. Frequency Allocations

Originally no exclusive frequency bands were available for satellites in the rapidly crowding spectrum below 10 GHz, but because of the availability of proven techniques with terrestrial equipment in the 4 and 6 GHz bands it was natural that these bands should be initially used for satellite communications and shared with terrestrial systems. But the clearly dominant consideration for future spectrum utilization is the advent of the communication satellite. In 1966, the existing technology made the satellite use of frequencies to about 15 GHz feasible and had the potential of extending that range by an order of

[90]Dale Hatfield, "A General Analysis of Domestic Satellite Orbit/Spectrum Utilization" (Washington, D.C.: U.S. Department of Commerce, Office of Telecommunications, PB 207397, December 13, 1971), p. 9.

magnitude.[91]

Using the 4 and 6 GHz bands with approximately three degrees of orbital separation between satellites and ten degree minimum elevation angles, some 24 satellites could be accommodated each using the total band for up and down transmission interference. Each satellite could have up to twenty-four 40 megahertz (40 million cycles per second) channels, each capable of up to 1,200 voice circuits or one television circuit per channel. But one of the most important parameters in determining minimum orbital spacing is Earth station antenna size. Antennas for such satellite systems would have to be on the order of one hundred feet in diameter for acceptable performance. At higher frequencies, particularly those above 10 GHz interference is less likely to be a problem with terrestrial systems and trade-offs can be made betweeen the size and space parameters of a satellite system.[92]

Because of ionospheric effects and high noise levels, the low limit on frequencies for use in satellite links is around 70 megahertz (MHz). Until about 10 GHz,

---

[91]Report by the Telecommunication Science Panel of the Commerce Technical Advisory Board, "Electromagnetic Spectrum Utilization--The Silent Crisis" (Washington, D.C.: U.S. Department of Commerce, October, 1966), p. 10.

[92]For additional information on this subject, see D. M. Jansky and M. C. Jeruchim, "Technical Factors and Criteria Affecting Geostationary Orbit Utilization." Communication Satellites for the 70's (Cambridge, Massachusetts: The MIT Press, 1971).

transmission is relatively free and then additional path loss, caused by rain, clouds, or fog, begins to reduce efficient transmission. Higher powered satellites and highly directive antenna systems can be used to overcome some path loss problems. Modern solar panel arrays can provide a satellite with up to five kilowatts of power for operation if necessary.[93]

Hypothetical systems have emerged from studies by Bell and others which have projected the possibilities for future satellite systems. Labeling spectrum and orbit space as "precious and limited resources which must be conserved", a system using frequencies in the 20 and 30 GHz bands was "designed" that used 50 satellites and 50 Earth stations and could offer up to 100 million voice circuits or equivalent. Each satellite weighed about five tons, used digital technology and had a total capacity of about four million voice circuits. Such systems far surpass today's needs but future telecommunications requirements may require such systems to be developed.[94]

. . . .

Except for the brief period of time around the turn of the century there was little if any true

---

[93]A more detailed discussion is provided in Advanced Communication Systems, pp. 199-227.

[94]Martin, Future Developments in Telecommunications, pp. 240-243.

competition among the common carriers for residential telephone and long distance service. However, technological advances have since introduced important competitive elements into the communications industry. Although regulation of the carriers appeared at first to be only an afterthought in an attempt to control the AT&T monopoly, the Commission more recently has been working hard to change its image and to take a more positive role in the regulatory process.

By adopting its competitive attitude in 1959, the FCC chose not to leave initiative for the services to the public that the new technologies could provide up to the established carriers. By authorizing private ownership of microwave systems the Commission only increased its workload and gained the disfavor of the established carriers. The industry's structure was beginning to change with technology and so was the FCC; it was attempting to serve the public interest.

The FCC was primarily established to insure that there was equity, order and efficiency in the assignment of the radio frequency spectrum. Technological advances in the use of this spectrum after World War II posed no major problem for the Commission, for Congress had given the agency, through the broad language of the 1934 Communications Act, the leeway it needed to keep step with technology.

In making the Above 890 decision, the FCC

established the fact that it recognized the assistance that technology could furnish to them in promoting the public interest. The Commission embarked on a policy of increased competition on its own initiative and challenged the long-standing natural monopoly thesis. As economist William G. Shepherd has observed:

> If the technology of telecommunications prescribed a natural monopoly, total and integrated control by one system under appropriate public constraints might be economically justified. Yet the technological margins between "necessity" monopoly and possible competition have been in a state of flux, particularly in the last two decades and the newer "operational" services have tended to be both more lucrative and more capable of supporting a more competitive market structure.[95]

Although AT&T, through the Bell Laboratories, is undoubtedly the Nation's leading contributor of telecommunications technology, its monopoly structure and financial policies appear to be ideally suited to impeding technical innovation. But market demands and regulation have promoted innovation and have minimized the negative effects of these factors to some degree. The telecommunications technology that has been available has seemed to always extend far beyond projected near-time communications requirements. The telecommunications policy makers who confronted the communication

---

[95]Shepherd, "The Competitive Margin in Communications," p. 95.

satellite issue were faced with the same public interest considerations that any group would have had in opening a new market that was based on a new technology. But such considerations by the FCC for domestic satellites would certainly have many factors; the impact of DOMSAT on existing market structures and the established carriers would be only one of the many that the Commission would have to consider.

CHAPTER II

THE PRECEDENTS OF DOMSAT

In the development of telecommunications policy, a relationship equally as important as that of Commission-to-carrier is that of Commission-to-Congress. The FCC was established by Congress both as an independent regulatory commission and as "an arm of the Congress" and to Congress, this relationship may mean independence from White House domination, but not necessarily independence from its Congressional parent.[1]

Congress made a major amendment to the Communications Act of 1934 with the passage of the Communications Satellite Act of 1962 and expanded the FCC responsibilities. *Sputnik I* had helped Congress to recognize that the commercial utilization of space could promote a wide range of benefits for the public. This could be accomplished through either the economic improvement of existing concepts or through the processes of technical innovation. Among all the projected commercial uses of space, communications

---

[1] Erwin G. Krasnow and Lawrence D. Longley, *The Politics of Broadcast Regulation* (New York: St. Martin's Press, Inc., 1973), p. 54.

was the one which took the strongest foothold and offered the greatest potential.[2] During the five years that elapsed between the launching of Sputnik I and the passage of this legislation which established the Communications Satellite Corporation (COMSAT), the questions of competition, ownership, operation, markets and boundaries were all addressed to some degree by a variety of parochial interests and activities.

The Corporation's creation provided policy foundations that were examined and challenged during the development of the DOMSAT policy (as discussed in Chapter III). This makes the understanding of the functions of the Commission and Congress in this area of telecommunications and the rationale for their actions important from the onset. In the 1960's, COMSAT's relationships with Congress, the FCC and the carriers were unique and added a level of complexity to the rapidly changing environment of the Commission and to its regulatory functions. Although DOMSAT compounded these complex relationships again ten years later, the foundations had been laid by the Commission and Congress in 1962 with COMSAT.

A. The Communications Satellite Act of 1962

By means of the Communications Satellite Act

[2]Hans K. Ziegler, "Space Communications--A Major Candidate for Commercial Utilization of Space," Advances in the Astronautical Sciences, 28 (1968), p. 91.

of 1962, Congress created the Communications Satellite Corporation (COMSAT). This was a public corporation--half owned by the major communications companies and half owned by individual investors--established to develop a commercial, international communications system using satellites, put it into operation and manage it in cooperation with foreign countries. The advent of communications satellite technology, the aspirations of individual companies in exploiting it, and public policies had brought about important changes in the structure of the U.S. overseas communications system. As in 1927 with the case of high frequency radio,[3] the government was anxious to promote the fastest possible development of the new communications technology [as well as an improved world leader image].[4]

1. Congressional Hearings

As would be true with any complex piece of legislation, the process of its enactment was not simple. Not only had difficult questions of ownership

---

[3]See Chapter I, note 72.

[4]Kurt Borchardt, Structure and Performance of the U.S. Communications Industry (Boston: Harvard University Press, 1970), p. 49, 53. When the Soviets launched Sputnik I, the U.S. had suffered a serious blow to its image as the scientific and technological leader of the world. The challenge not only affected the public's image of U.S. military superiority but also the ability of the Republican administration to conduct foreign policy.

been raised but also there had been a change in administrations by the time the issues had reached their full intensity. To complicate matters, there was no agreement on an ownership policy for commercial communications satellites either within the communications industry or the Congress.

In 1961, the Senate Subcommittee on Monopoly held hearings into the pros and cons of existing government policies and established organizations for space communications; so did the House Interstate and Foreign Commerce Committee and the House Committee on Science and Astronautics. More than eighteen months passed between Eisenhower's first statement of policy and the Kennedy legislation being signed into law. During that period, the FCC was the first to face the issues that were raised.

a. The Ownership Question

The alternatives of ownership for commercial communications satellites were basically (1) government ownership, (2) carrier ownership, and (3) private, broad-based ownership.[5] Congressional interest was soaring. Between June 14 and August 24, five Congressional committees held 21 days of hearings on

---

[5]Jonathan F. Galloway, The Politics and Technology of Satellite Communications (Lexington, Massachusetts: D.C. Heath and Company, 1972), pp. 47-48.

communications satellites despite the fact that there was no legislation pending on the subject.[6]

At the same time, the FCC initiated a formal Notice of Inquiry addressed to the question of ownership and operation of such a venture, specifically soliciting the views of industry as to what plan of participation was considered best. Twelve interested parties responded and there was some agreement for joint ownership and operation of the system.[7] The options were being filtered through the political and psychological climate of the day.

The overseas carriers argued that potential economies of scale would be effected by treating satellites as an extension of existing submarine facilities. They proposed a joint venture whereby satellite ownership would be assigned exclusively to them. The aerospace industry took an entirely different view. General Electric and Lockheed, in particular, called for the creation of a carrier's carrier and argued that the entity's ownership should

---

[6]Horace P. Moulton, "Communications Satellites--the Proposed Communications Satellite Act of 1962," Business Lawyer, 18 (November 1962), p. 175. See Galloway, note 5, supra, p. 11, for a breakout of the Congressional Committees dealing in communications.

[7]U.S. Congress, House, Committee on Science and Astronautics, Commercial Applications of Space Communications Systems, H. Rept. No. 1279, Oct. 11, 1961, 87th Cong., 1st Session, pp. 23-24, and FCC Docket No. 14024, Notice of Inquiry, April 3, 1961. See Chapter III, note 22.

include equipment suppliers and the public at large as well as the overseas carriers.[8]

The response of all common carriers, domestic as well as international, generally expressed opposition to participation in ownership by noncarriers. As AT&T put it, such arrangements would enable hardware suppliers, who have no responsibility to the public for quality or scope of service, to influence the common carriers' future undertakings.[9]

AT&T Vice-President James E. Dingman testified before the Senate that communications satellites were really "no big breakthrough"; they would not make undersea cables obsolete and they certainly had no potential for domestic use. However, the carriers were still sincere and enthusiastic in their desire to help advance satellite communication [the Nation needs more public spirit like that]. He stated:

> This position may be construed by some as stemming from the selfish interests of my company which is the largest of the carriers involved [it's the largest of ALL carriers!]. Let me assure you that it is not.
>
> Let one thing be crystal clear: AT&T has no desire or intention of seeking to control the communications satellite

---

[8]Manley R. Irwin, <u>The Telecommunications Industry--Integration vs. Competition</u> (New York: Praeger Publishing, 1971), pp. 97-98.

[9]Lloyd D. Musolf, <u>Communications Satellites in Political Orbit</u> (San Francisco: Chandler Publishing Co., 1968), p. 25.

> system to its competitive advantage. . . . Hard as it may be for some to understand, our sole interest is in the earliest practicable establishment of a world-wide commercial satellite system useful to all international communications carriers and agencies both here and abroad.[10]

The Justice Department neither suggested nor endorsed any specific plan, but instead specified four conditions necessary for joint ventures in order that they be consistent with antitrust considerations:

> 1. All interested communications common carriers be given an opportunity to participate in ownership of the system.
>
> 2. All interested communications common carriers be given unrestricted use (on nondiscriminatory terms) of the facilities of the system whether or not they elect to participate in ownership.
>
> 3. All interested parties engaged in the production and sale of communications and related equipment be given an opportunity to participate in ownership of the system.
>
> 4. All interested parties engaged in the production and sale of communications and related equipment be given unrestricted opportunity to furnish such equipment to the system whether or not they elect to participate in ownership.[11]

By reporting on its Notice of Inquiry, the FCC was the first agency to confront the policy choices,

---

[10]Michael E. Kinsley, Outer Space and Inner Sanctums: Government, Business and Satellite Communications (New York: John Wiley and Sons, 1976), pp. 9-10. Bracketed commentary added. Dingman's testimony was before the Committee on Aeronautical and Space Sciences on March 6, 1962.

[11]Irwin, The Telecommunications Industry, pp. 99-100.

and it must be noted that it acted with unusual dispatch. In its report of May 24, 1961, it stated:

> We fail to see why ownership or participation by the aerospace industry in the communications industry would be beneficial or necessary to the establishment of a satellite communications system to be used by the common carrier industry.[12]

With this observation the Commission rejected GE's plan to establish a satellite corporation, and placed the Justice Department and the Assistant Attorney General, Lee Loevinger, in a dilemma by failing to support a joint ownership policy. However, in apparent deference to the FCC, Justice modified its requirement for aerospace "ownership" rights, and substituted "participation" as its guideline for a satellite venture.[13]

b. The Kennedy Administration Bill

The voice of President Eisenhower had been one of the first heard on the subject of commercial satellite communications:

> The commercial application of communication satellites, hopefully within the next several years, will bring the nations of the world closer together in peaceful relationships as a product of this Nation's program of space exploration. . . . The Nation has traditionally followed a policy of conducting international telephone,

---

[12]Ibid., p. 99.

[13]U.S. Congress, Commercial Applications of Space Communications Systems, p. 26.

> telegraph and other communications services through private enterprise subject to government licensing and regulation. We have achieved communications facilities second to none among the nations of the world. Accordingly, the Government should aggressively encourage private enterprise in the establishment and operation of satellite relays for revenue-producing purposes.[14]

But by the Fall of 1961, the Washington environment had changed. The Kennedy Administration was now in the White House and the COMSAT controversy was fully monopolizing Congress.

President Kennedy viewed Eisenhower's policy as "turning control of space communications over to AT&T"[15] and on July 24, 1961 had announced that a policy of private ownership and operation of the U.S. portion of the system was favored provided that such ownership and operation met the following policy requirements:

1. New and expanded international communications services be made available at the earliest practicable date;

2. Make the system global in coverage so as to provide efficient communication service throughout the whole world as soon as technically feasible, including service where individual portions of the coverage are not profitable;

---

[14]U.S. Congress, Senate, Committee on Aeronautics and Space Sciences, Staff Report, Documents on International Aspects of the Exploration and Use of Outer Space, S. Doc. No. 18, May 9, 1963. From a "Statement by President Eisenhower on the Commercial Use of Communication Satellites," December 31, 1959.

[15]Galloway, The Politics and Technology of Satellite Communications, p. 47.

3. Provide opportunities for foreign participation through ownership or otherwise, in the communications satellite system;

4. Nondiscriminatory use of, and equitable access to, the system by present and future communication carriers;

5. Effective competition, such as competitive bidding, in the acquisition of equipment used in the system;

6. Structure of ownership or control which will assure maximum possible competition;

7. Full compliance with antitrust legislation and with the regulatory controls of the Government;

8. Development of an economic system, the benefits of which will be reflected in oversea communication rates.[16]

The Executive Secretary of the National Aeronautics and Space Council, Mr. E. C. Welsh, was tasked to prepare a coordinated draft proposal for translating Kennedy's policy into effective legislation. By January 1962, after many meetings of the Council, constructive language evolved. The Administration's bill (H.R. 10115 or S. 2814) provided for the establishment, ownership, operation and regulation of a commercial communications satellite system and authorized the creation of a "privately owned and profit-operated Corporation [COMSAT]." COMSAT was to be financed from the sale of securities to the public which included, but was not limited to, common carriers or otherwise chosen companies or

---

[16] U.S. Congress, Commercial Applications of Space Communications Systems, pp. 25-27.

individuals. It would not be an agency or establishment of the U.S. Government but it would be subject to the pertinent provisions of the Communications Act of 1934, as amended, and of the District of Columbia Business Corporation Act.[17]

2. H.R. 11040 Becomes Law

Opinion in Congress was now oscillating between the two extremes of government and carrier ownership and private ownership was seen by some as a violation of antitrust laws and a giant giveaway of government investments in communications satellite technology.[18] In an August 1961 letter to the President, the liberal Democrats in Congress (three Senators and thirty-two Representatives) had urged that a hasty decision on the space communications issue not be made in order that the general "national interest" might be determined. However, there was still no agreement within Congress as to what the national interest was nor how it could best be determined or served.

---

[17]Musolf, Communications Satellites in Political Orbit, pp. 57-58.

[18]Galloway, The Politics and Technology of Satellite Communications, p. 52. Senator Kefauver, Chairman of the Subcommittee on Antitrust and Monopoly, led the opposition. Assistant Attorney General Loevinger noted that the technology was made possible through government-sponsored R&D mainly because there was national interest in the establishment of a satellite communications system.

a. Opposing Views

No fewer than ten bills on the subject were bouncing around Congress in 1962. In a simplified picture, the cast of characters looked like this. There was Senator Kerr of Oklahoma who favored private ownership with minimal government regulation, Senator Kefauver of Tennessee, who favored government ownership, at least initially, and Senator Pastore of Rhode Island, who wanted private ownership with strong government control specified in the enabling legislation. The President's proposal had been introduced to both houses on February 7, and questions concerning the role of the Executive and the bill's domestic and foreign policy implications were also causing debate.[19]

The first committee report on S. 2814, the President's legislation, was issued on April 2 by the Senate Space Committee. In the House on the same day, Congressman Oren Harris introduced H.R. 11040 which was identical to S. 2814 as amended by the Senate Space Committee. With minor refinements, H.R. 11040 was passed in the House on May 3 by a vote of 354 to 9. It was then sent to the Senate, where it was referred to the Commerce Committee.[20] Senate

---

[19]Ibid., pp. 52-53.

[20]Ibid., pp. 64-65. See also note 29, *infra*.

activity continued independently on S. 2814. Changes were made by the Committee on Commerce which would restrain the monopoly and protect the taxpayers to a far greater extent than what had been previously proposed.[21] Senator Pastore was especially concerned that domination by one communications common carrier (AT&T) should be avoided.[22] The committee amended subsection 102(c) to express the intent of Congress regarding Federal antitrust laws and 102(d) _so that nothing in the act could preclude the use of such [COMSAT] systems for domestic communications services_ where consistent with the provisions of the act.[23]

---

[21]Musolf, _Communications Satellites in Political Orbit_, p. 54.

[22]Galloway, _The Politics and Technology of Satellite Communications_, p. 57.

[23]U.S. Congress, Senate, Committee on Commerce, _Communications Satellite Act of 1962_, S. Rept. No. 1584, June 11, 1962, 87th Congress, 2nd Session, pp. 13-14, emphasis added. Section 102(c) was modified to dispel fears that the legislation might be construed as a total exemption of the corporation and participants from antitrust laws. However, this action could also be considered as an affirmation of the proposition that competition, however qualified, is presumptively the national policy, and that any exemption found in the act should be narrowly construed. See also Legislative Note, "The Communications Satellite Act of 1962", Chapter III, note 44, p. 397. The clarification of subsection (d) was made to avoid any possible inference that may be drawn from the other provisions of the bill that Congress had made a policy determination that use of the system be limited to international communications. However, the committee felt that it was unlikely that the system would not be useable for domestic services (because of technical and economic limitations) but that they were entirely consistent with the Act.

Debate in the Senate was turned into a strategy of filibuster by the bill's opponents.[24] As a consequence, cloture was imposed on August 14 (the first time it had been successfully used since 1927) to end debate and on August 17, the bill, which was in essence H.R. 11040 with everything after the enacting clause eliminated and the body of S. 2814 (as amended by the Commerce Committee) inserted in lieu thereof, finally passed the Senate and was sent to the House.[25] The bill won final House approval on August 27, 1962.

b. The Final Act

When President Kennedy signed the Communications Satellite Act of 1962 on August 31, one of the most controversial pieces of legislation of the 87th Congress became law and the opponents of COMSAT were finally defeated.[26]

The purpose of the Act is best summarized by

---

[24]Primary opposition were the supporters of a government-owned approach; Kefauver (D., Tenn.), Morse (D., Ore.), Yarborough (D., Tex.), Gore (D., Tenn.), Long (D., La.), Burdick (D., N.D.), and Neuberger (D., Ore.).

[25]Musolf, Communications Satellites in Political Orbit, pp. 57, 97-108. The bill that had left the House over three months earlier was hardly the same bill that was just returned. However, it was recognized that any amendments to the Senate version could be subjected to filibuster again.

[26]Galloway, The Politics and Technology of Satellite Communications, p. 69.

Sections 102(a) and (b) of the Act:

> (a) The Congress hereby declares that it is the policy of the United States to establish, in conjunction and in cooperation with other countries, as expeditiously as practicable a commercial communications satellite system, as part of an improved global communications network, which will be responsive to public needs and national objectives, which will serve the communication needs of the United States and other countries, and which will contribute to world peace and understanding.
>
> (b) The new and expanded telecommunications services are to be made available as promptly as possible and are to be extended to provide global coverage at the earliest practicable date. In effectuating this program, care and attention will be directed toward providing such services to economically less developed countries and areas as well as those more highly developed, toward efficient and economical use of the electromagnetic frequency spectrum, and toward the reflection of the benefits of this new technology in both quality of services and charges for such services.[27]

With respect to the Communications Act of 1934, the COMSAT Act states that the corporation that was created by the Act [Communications Satellite Corporation] shall be fully subject to the provisions of the Communications Act. However it further states that:

> Whenever the application of the provisions of this Act shall be

[27]The Communications Satellite Act of 1962, Pub. L. No. 87-624, 87th Congress, 2nd Session (August 31, 1962), 76 Stat. 419. Emphasis added.

> inconsistent with the application of the provisions of the Communications Act, the provisions of this Act shall apply.[28]

In creating COMSAT as a joint venture, subject to Government influence but owned and operated by broad-based private interests, Congress rejected a number of alternatives such as completely governmental projects (like the Atomic Energy Commission or the Tennessee Valley Authority), purely commercial joint ventures, and single-company operations.[29] COMSAT, like the FCC, is a creature of Congress, but not by accident. It was not created because "no entry would otherwise take place." The Government or AT&T could have acted alone or separate companies could have established individual segments of a global relay.[30]

Rapid development was a strong consideration (Kennedy's criteria--"at the earliest practicable date"--was in partial response to a projected deficiency in international communications capability

---

[28] Ibid., Section 401.

[29] Harvey J. Levin, "Organization and Control of Communications Satellites," University of Pennsylvania Law Review, 113 (January, 1965), p. 324.

[30] AT&T was obviously in the best position to develop the system. Success with Telstar (a wholly Bell-sponsored random-orbit communications satellite which was launched in July 1962) indicated both capacity and desire for vigorous independent action. To this could be added the company's great financial resources, which might have been considerably augmented by such foreign participants as the British Post Office.

and to meet the alleged requirements of national prestige in the "cold war"), probably stronger than commercial considerations would have dictated. If time had been of no concern, the country might have waited until the market could support multiple independent private systems or joint ventures limited to parties without vested communications interests.[31] On the basis of costs, single-company ownership would have been easily possible, especially if NASA had charged no more than marginal costs for launching and tracking. Despite these considerations, opposition to a single-company ownership was overwhelming in view of the threat of monopoly, accompanied by anti-trust and regulatory problems. Single-company ownership, in fact, was never _formally_ proposed in Congress.[32]

The remedy which Congress finally selected was thus obviously not Commission regulation pure and simple. It was instead a set of special techniques intended to produce, by internal organizational constraints, some of the results that a competitive economic structure would have produced externally.[33]

---

[31]Levin, "Organization and Control of Communications Satellites," pp. 325, 356.

[32]Ibid., pp. 332-333.

[33]Ibid., p. 338.

B. Space Age Regulation

The burden of satellite communications regulation falls primarily on the FCC. The COMSAT legislation imposed elaborate direct controls by the Commission on this "common carrier's common carrier", more comprehensive and more complex than any of the regulatory apparatus that had been used previously for the supervision of traditional communications carriers. In its expanded role, the Commission could require additional facilities from COMSAT if called for by the public interest. It could authorize construction, operation and ownership of ground terminal stations of the system by the Corporation, or by private communications carriers, or the two jointly. In general, the FCC was empowered to "make rules and regulations to carry out the provisions of this Act."[34]

The Commission now had an additional opportunity to expand on its competitive communications policy. It was evident to the drafters of the legislation that the new COMSAT Corporation would have to consider the many public and national interest considerations inherent in this new area of endeavor. Consequently to insure that all interests were faithfully considered, Congress had applied a scheme of regulation that was literally

[34] "Communications Satellite Act of 1962," Harvard Law Review, 76 (1962), pp. 390-391. See also Levin, "Organization and Control of Communications Satellites," p. 316.

unprecedented.[35] An example of the Commission's added duties in satellite matters is typified by the following excerpt from the Act which required the Commission to:

> insure effective competition, including the use of competitive bidding, where appropriate, in the procurement by the Corporation and communications common carriers of apparatus, equipment, and services for the establishment and operation of the communications satellite system and satellite terminal stations.[36]

In the exercise of its new authority, the Commission was quickly confronted with essentially three new considerations which surfaced. The use of outer space for communications had international ramifications that required some rethinking of the traditional international frequency allocation process that had developed through the efforts of the International Telecommunications Union (ITU) and its predecessors since 1903. In addition, this new technology offered cost and performance advantages that were attractive to both the carriers and businesses

---

[35] Rosel H. Hyde, "Space Age Regulation," Presentation before the Midwest Association of Railroad and Utility Commissioners (Chicago, Illinois: July 7, 1964), p. 5. The COMSAT Act includes provisions where the Commission could actually direct COMSAT to establish communications to a particular part of the world, regardless of whether it was profitable or not, if the Secretary of State advised that it was in the national interest.

[36] The Communications Satellite Act of 1962, Pub. L. No. 87-624.

alike and questions of who was authorized to use the COMSAT system were quickly raised. Finally, in 1965, these questions gave birth to the idea of applying satellite communications technology to domestic communications applications and the domestic satellite policy issue was placed before the Commission.

1. International Considerations

Satellites, which are oblivious to national borders and physical obstructions such as mountains, oceans and great distances, make the distinction between domestic and international communications a purely artificial one.[37] As the era of space communications progresses, it is important to recognize its impact on other nations of the world.[38]

Just as nations feel that a stockpile of weaponry is imperative for security and prestige, so too, is it believed that a domestic satellite link for communications is a guarantee for independence and status.[39] The power elites of the developing countries

---

[37] Kinsley, Outer Space and Inner Sanctums, p. 137.

[38] Neil P. Hurley, "Satellite Communications," America, 115 (August 27, 1966), p. 204.

[39] Ibid., p. 206. Although it was the express wish of President Kennedy that the envisioned system of satellite communications be truly global, "national" motives represent the most difficult obstacle to this and it's hard to see how a proliferation of satellite systems can be avoided.

are eager for the communications power that satellites might help to provide, consolidating national power and promoting a sense of national unity and loyalty.[40] Although the considerations imposed by nationalism are becoming more pertinent in today's international arenas with the emergence of each new nation, this thought is only identified here as an international consideration to be addressed by the determiners of future telecommunications policy in forums like the ITU.

The basic questions that were addressed by the FCC in its communications satellite policy decisions were those that evolved from the use of international resources for the development of this technology. These decisions were important since they represented initial policy. It is reasonable to believe that future generations of all nations are either influenced or constrained considerably by these policies or by whatever systems of satellite communications or related systems of international policy that result. The question of geostationary orbit utilization represented a unique factor for consideration in the FCC policy-making process.

a. <u>The International Telecommunication Union</u>

The creation of the International Telegraph

---

[40] O. W. Riegel, "Communications by Satellite: The Political Barriers," <u>Quarterly Review of Economics and Business</u>, <u>11</u> (1971), p. 34.

Union, the predecessor of the ITU, was the first important step toward removing artificial barriers to communications; it paved the way for long-distance exchange of messages without regard to political boundaries. Satellite communications represent a vast improvement in this exchange of messages both within and among nations and at the same time create new problems of regulation and operation which were unimagined when the ITU began.[41]

Dr. George A. Codding, Jr., a recognized authority on the ITU, recently noted to the Fourth Telecommunications for Government Conference:

> The ITU performs an indispensable task in allocating the radio frequency spectrum to various telecommunications services; it performs an essential task in the adoption of measures for the safety of life; it performs a necessary task in registering frequencies; and it performs a valuable task in bringing together telecommunications experts to work together to solve common problems and, in general, to share their expertise.[42]

The ITU is considered by some to be no weaker internationally than are national agencies like the FCC, when

---

[41]Gordon L. Weil, Communicating by Satellite: An International Discussion (New York: The Twentieth Century Fund, 1969), p. 3. The ITU is the oldest specialized agency within the United Nations. Founded in 1865 to facilitate international communications, its membership is more than 146 nations.

[42]George A. Codding, Jr., "The U.S. and the ITU in a Changing World," OT Special Publication 75-6, Department of Commerce, December 1975, p. 9.

it comes to making long-range communications policies. There appears to be no alternative to the ITU as there is no acceptable alternative to orderly planning for the use of satellites throughout the world.[43]

Since the satellite's inception, those concerned with communications have been aware that only through international cooperation can the tremendous potential of this technology's application to communications be realized. Also recognized was the need for some international organization to coordinate and regulate the several communications satellite systems that would become operational throughout the world; the ITU was the logical candidate for this task.[44]

Briefly stated, the structure of the ITU comprises a Plenipotentiary Conference composed of all member countries and meeting approximately every seven years to revise the International Telecommunication Convention, an Administrative Council that directs the ITU's affairs between Plenipotentiary Conferences, two technical International Consultative Committees (CCI's), one for radio (CCIR) and one for telephone and telegraph (CCITT), a General Secretariat with an elected Secretary General and an International Frequency

---

[43]Paul L. Laskin, Planning for a Planet, An International Discussion on the Structure of Satellite Communications (New York: The Twentieth Century Fund, 1971), p. 24.

[44]Ibid., pp. v-vi.

Registration Board (IFRB).

The IFRB was of primary importance to U.S. satellite communications considerations. It is composed of five elected members, with its own specialized secretariat, that receives notification from countries seeking to use a particular radio frequency; the IFRB maintains a Master Register of frequencies, and if the proposed use conforms to the regulations, gives it a legitimate status by entering it on its Register. In 1934, the vice-director of the International Telecommunication Union Bureau had defined the right of priority of an administration to a frequency as moral only and not a juridicial priority.[45] However, in practice administrations have generally observed and acted upon the ITU's notification procedure of the IFRB.

It had not be necessary to raise the problem of allocation of frequencies for space services prior to the 1959 ITU Conference.[46] It was then that the United States first proposed the assignment of ten frequency bands to accommodate the function aspects of space communications. Only interim allocations

---

[45]George A. Codding, Jr., The International Telecommunication Union (Leiden: E. J. Brill, 1952), p. 190.

[46]The space technology itself did not exist prior to 1956. See Werner, "A Lawyer Looks at Our Communications Policy," Chapter III, note 99.

were made at that time for research purposes, but the general matter was later quite adequately resolved by permanent allocations at the 1963 Extraordinary Administrative Radio Conference in Geneva.[47]

By accepting the responsibility for negotiating frequency allocations for space and earth-space services, the ITU extended the scope of its law-making treaties from airspace to outer space. The development of communication satellites was instrumental in creating new dimensions to the already complex international legal and technical problems. The 1963 Conference assigned to the IFRB the responsibility of the notification and recording of frequency assignments to space satellite stations and to the Earth stations with which they are in communication.[48]

Spectrum management is not a static affair relating to tangible objects to which legal rights and obligations can be attached. It is dynamic by nature and calls for flexible administrative measures to achieve coordination and to avoid harmful interference.[49] Since there are variations in radio wave

---

[47]Galloway, The Politics and Technology of Satellite Communications, pp. 18, 19, 78.

[48]Erin Bain Jones, Earth Satellite Telecommunications Systems and International Law (Austin, Texas: The Encino Press, 1970), pp. 96-97.

[49]Charles H. Alexandrowicz, The Law of Global Communications (New York: Columbia University Press, 1971), p. 33.

length propagation characteristics, only waves of certain lengths can pass through the atmosphere and ionosphere. For this reason only selected frequencies within the radio spectrum can be used for space communications. Early frequency assignments required the sharing of channels with terrestrial services.

However, the 1971 World Administrative Radio Conference on Space Telecommunications (WARC-ST) made new options possible.[50] An eight-fold expansion in the radio spectrum for point-to-point satellite communications, suitable for use during the 1970's was allocated by the 1971 WARC-ST to be effective January 1, 1973. This announcement [in August 1971] certainly was an important factor in the final DOMSAT decision. For the first time millimeter-wave frequencies (above 30 gigahertz) were available and frequency allocations were made for satellites in the bands above 10 GHz.[51] Although in the higher bands, signals are heavily absorbed due to clouds, rainfall and the atmosphere itself, the bandwidths available are much higher which permits satellites of very high channel capacity to be constructed.[52]

---

[50]Jones, Earth Satellite Telecommunications Systems and International Law, p. 97.

[51]Allocations prior to 1963 were in the 4-6 gigahertz range only.

[52]Philip J. Klass, "Satellite Radio Spectrum Expanded," Aviation Week and Space Technology, 95

The 1971 WARC-ST specifically allocated [at the urging of the U.S.] frequencies in the 11.7-12.2 gigahertz range for domestic-regional satellites in the Western Hemisphere.[53] But allocations by region [although workable for terrestrial systems] become outmoded in the face of the extensive coverage of satellites--far beyond the confines of any one region. Rapid technological change as well as the considerable increase in the number of sovereign nations involved in international communications and the ITU have added complications to the international frequency allocation process which could not have been foreseen when the ITU's structure was first established.[54]

b. Orbital Parking Slots

Concern has been expressed at many levels over

(August 2, 1971), p. 14. Much of the required technology needed to use these frequencies was then available or soon to be. Four of the eight original proposals submitted for DOMSAT systems included these frequencies as well as the then existing 4-6 gigahertz band. The allocation was slightly higher than had been proposed by the U.S. but was expected to be superior because of reduced interference problems with existing terrestrial services. This consequently required slight modification to those DOMSAT proposals. The SBS proposal, which would not be filed until December 22, 1975, would also plan to use the higher frequencies.

[53]"U.S. Gets Space It Wanted," Broadcasting, 81 (July 26, 1971), p. 44. This is Region 2. For a detailed discussion of the 1971 WARC-ST allocations, see "The World Administrative Radio Conference for Space Telecommunications, Geneva, 7 June-17 July 1971," Telecommunications Journal, 38 (October, 1971), pp. 673-682.

[54]Weil, Communicating by Satellite, pp. 6, 8.

the sufficiency of the supply of orbital slots over the equator to accommodate the present and future requirements for communications satellites in the American hemisphere. The limits of the spectrum are known, thus making it a finite resource that must be allocated carefully. While space in comparison may seem limitless, the development of the geostationary satellite has quantified it to some degree.[55] For example, the service arc for U.S. continental coverage is approximately 70° wide (based on elevation angles of greater than 10°). With 3° of orbital spacing to minimize frequency interference, the 24 orbit locations could be filled with a total space segment investment of less than a billion dollars.[56]

Orbital slots have been analogized to land on a frontier, waiting to be preempted by the first comer; but squatters rights are not the answer as a look at the frequency allocation process helps to explain. It is not possible to establish legal ownership of a frequency assignment for a nation, a corporation, or an individual. Under international law expressed through the ITU for more than half a century, the

---

[55] Ibid., pp. 8-10.

[56] Emeric Podraczky, "Utilization of the Geostationary Satellite Orbit," Telecommunications (January, 1975), p. 30. This cost is trivial in comparison to the huge investment that presently exists in terrestrial systems.

radio spectrum is a world resource. It must be said, however, that the rule in frequency allocation has tended to work out as "first come; first served." Nevertheless, the ITU has never sanctioned the principle as a matter of law.[57]

The logic of radio frequency allocation involves:

1. The allocation of a portion of the spectrum and determination of minimum engineering standards for the desired service;

2. Determination of engineering standards for specific systems;

3. Determination of the allocation of frequency assignments with respect to the geographic space to be served by the signals; and

4. Determination of which organization should conduct the radio service at each of the specific "frequency assignments" produced in Step 3 (this is often referred to as the licensing of the operator).

The four steps in the process are all interrelated logically, technically, administratively and politically. The "parking slot" problem involves Steps 2, 3 and 4. More naive formulations seem to assume that Step 4 is readily available for the first comer. In reality it is much more complex and all steps are necessary. It is illusion to fancy that private property "slots"

---

[57] Dallas W. Smythe, "The 'Orbital Parking Slot' Syndrome and Radio Frequency Management," Quarterly Review of Economics and Business, 12 (1972), p. 7. See also G. A. Codding, Jr., Broadcasting Without Barriers (Paris: UNESCO, 1959).

(like homesteads) can be "staked out" on the orbital arc without some form of bargaining between interest groups being involved.[58]

In the United Nations, the Declaration of Legal Principles immunizes outer space and the celestial bodies against appropriation by claim of sovereignty, by means of use or occupation, or by any other means. In the exploration and use of outer space, states will be guided by the principle of cooperation. Articles I and II of the Declaration make it quite clear that national sovereignty is not extended from airspace to outer space. Consequently, the parking space concept is not consistent with the provisions of the Declaration.[59]

Source: Adapted from Podraczky, "Utilization of the Geostationary Satellite Orbit.

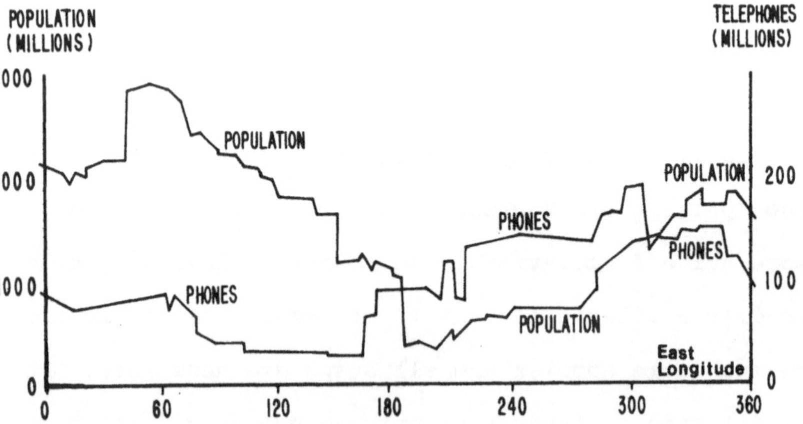

Figure 1. Population and Number of Telephones Seen From the Geostationary Satellite Orbit.

---

[58]Ibid., pp. 8-9, 13.

[59]Alexandrowicz, <u>The Law of Global Communications</u>, pp. 35-36.

Figure 1 gives a plot of the population as seen from the geostationary orbit. From this it is evident that every point on the orbit is of interest to a large number of people (300 million minimum). Also plotted is the number of telephone sets seen, which is considered to be a good indication of the voice traffic demand for satellite, although not necessarily the only one. When the geostationary orbit is viewed from this perspective, the distinction between domestic, regional or international service becomes somewhat arbitrary once again. For future systems it may be necessary to drop such distinctions for as satellites become more complex, their designs become more intimately related to the orbital positions to be used.[60]

The idea of efficiency in orbit utilization can be plagued with problems of interpretation. For example, a disaster and emergency warning system using small portable antennas may be inefficient in its use of the geostationary orbit and the spectrum but socially beneficial. Perhaps a video distribution system should not be compared to a heavy-route telephone trunk line for similar reasons.[61] Complicating the problem is the fact that there is no easy definition for "efficient

---

[60]Podraczky, "Utilization of the Geostationary Satellite Orbit," p. 31. A satellite looks at approximately 140° of arc on the earth's surface at the equator.

[61]Ibid., p. 30

orbit use" available. Not restricted to communication satellites only, the geostationary arc can also be used for other satellite systems which can provide services other than communications, such as weather tracking, resource mapping, navigation, and broadcasting. Trade-offs between the options have to be made.

2. Authorized Users

The Communications Satellite Act of 1962 made it quite clear that the Communications Satellite Corporation could not favor one carrier over another. For that very reason the Commission was required to:

> insure that all present and future authorized carriers shall have nondiscriminatory use of and equitable access to the communications satellite system and satellite terminal stations.[62]

When the FCC addressed the authorized user question, twenty-seven different interest groups filed comments.[63] The international carriers insisted that the intent of the 1962 Act had been to protect their investments by giving them exclusive rights to buy and resell COMSAT circuits. For example, International Telephone and Telegraph (ITT) argued that the carriers provided the service of transforming a "raw" satellite channel into a "useable circuit."

---

[62]Hyde, "Space Age Regulation," p. 6.

[63]This was in response to Docket No. 16058, In the Matter of Authorized Entities and Authorized Users under the Communications Satellite Act of 1962, 4 FCC 2d 421.

But users such as International Business Machines (IBM) replied that they would rather purchase raw channels directly from COMSAT.[64]

The Commission's decision ruled that COMSAT could furnish satellite services and channels only to other international common carriers except in "unique or exceptional circumstances."[65] While it declined to permit COMSAT to compete freely and directly with the carriers, it did order the carriers to reflect the large cost savings [made possible by the satellite] directly in their international rates. What was disappointingly lacking in the FCC's decision from a market perspective was any explicit recognition by the Commission that similar economic advantages could possibly have been obtained with a less restrictive ruling on COMSAT and that such a ruling, at the same time, could have freed COMSAT for effective competition with the carriers. The Commission was justified though in trying to see to it that the benefits of satellite technology were passed on equally to all customers of the common carriers, either large or small.[66]

---

[64]Kinsley, Outer Space and Inner Sanctums, pp. 50-51.

[65]4 FCC 2d 12, Public Notice 66-563.

[66]Alfred E. Kahn, The Economics of Regulation: Principles and Institutions, Vol. II (New York: John Wiley and Sons, 1971), pp. 227-233.

The Government was also required by the Commission ruling to go through the carriers for satellite service unless "national interest" dictated otherwise. The FCC was called upon by the carriers to define "national interest" more specifically when the carriers filed an appeal with the Commission over a contract that the Department of Defense (DOD) had awarded to COMSAT for circuits to the Far East. The FCC ruled in the carriers' favor and adopted a restrictive interpretation of COMSAT's authority.[67]

Two other decisions of note occurred in this period. The FCC ruled that satellite ground stations should be owned half by COMSAT and half by the carriers, reversing an earlier tentative decision of sole COMSAT ownership. Carrier interests were being well looked after but so were the interests of COMSAT.

When AT&T requested permission to build a new transoceanic cable between the U.S. and Spain, the Commission informed them that agreement would have to be established on the "proportionate fill" of both cable and satellite facilities in such a manner that both facilities reached 100 percent circuit utilization at approximately the same time.[68] This protection

---

[67] Thomas P. Murphy, "Technology and Political Change: The Public Interest Impact of COMSAT," The Review of Politics, 33 (July, 1971), p. 411.

[68] Ibid., p. 413.

of technology has striking similarities to the 1927 cable/radio competition[69] with good reason; innovation can not only have a major effect on the market structure but it can also have a significant effect on the carrier's rate base, undersea cable still being the more capital intensive technology.

3. The DOMSAT Question

The idea of using communication satellites for domestic applications was first pondered in 1964. The Hughes Aircraft Company was interested in developing markets for its satellites and as a consequence, Dr. Harold Rosen sold the American Broadcasting Company the concept of having a separate company satellite, a "national network in the sky", which would simultaneously bypass AT&T's expensive long-distance circuitry and COMSAT's unchallenged monopoly of sky-circuitry.[70] Potential Hughes' customers up until this point had

---

[69]See Chapter I, notes 2, 71 and 72.

[70]Herbert I. Schiller, "Communications Satellites: A New Institutional Setting," Bulletin of the Atomic Scientists (April, 1967), p. 7. The system would deliver network television programs. With ABC, "it is primarily a matter of economics" (i.e., comparable or better quality service at reduced rates. ABC is currently using Western Union's "Westar" for occasional service with substantial savings, as compared to AT&T terrestrial links. Letter from Mr. Anthony Cusumano, Director of Traffic, ABC, to Robert S. Magnant, May 21, 1976. Kinsley suggests that ABC's filing was partly a battle maneuver in the struggle to prevent the telephone company from raising its rates. See note 10, *supra*, p. 138.

been limited to the U.S. Government and COMSAT.[71]

In an application to the Commission dated September 21, 1965, ABC requested authorization for a satellite for television broadcast distribution. This made use of an inherent characteristic of communications satellites that has no exact terrestrial equivalent: the ability to deliver the same wideband message to many distant locations simultaneously. By "broadcasting" every message it receives over a wide geographic area, the satellite would provide multipoint distribution services via a single relay point.[72] An opposition to the application was filed by COMSAT and the request was returned to ABC by the FCC without prejudice pending the resolution of certain basic issues of public policy.[73] In its letter to ABC, the Commission stated:

> Your application proposed a use of space techniques which is outside the purview of the established rules of the Commission. Furthermore, the

---

[71]Charles E. Silberman, "The Little Bird That Casts a Big Shadow," Fortune, 75 (February, 1967), p. 111.

[72]Final Report, President's Task Force on Communications Policy, December 7, 1968 (Washington, D.C.: U.S. Government Printing Office, 1969), 0-351-636, Chapter 5, p. 4. The ALOHA Network also makes use of this feature. A second unique characteristic, dynamic reallocation of capacity, is to be utilized in the proposed SBS system (see Chapter IV).

[73]Walter R. Hinchman, "Public Policy and the Domestic Satellite Industry" (Washington, D.C.: U.S. Government Printing Office, 1973). A paper presented before the Interstate Commerce Commission Conference--1972.

> unique nature of the proposal presents basic questions of law and policy which must be resolved before a proposal such as yours could be considered.[74]

Since the Commission had a statutory responsibility to study new uses of radio and generally encourage the larger and more effective use of radio in the public interest, it issued a Notice of Inquiry on March 2, 1966; Docket No. 16495--In the Matter of the ESTABLISHMENT OF DOMESTIC NONCOMMON CARRIER COMMUNICATIONS SATELLITE FACILITIES BY NON-GOVERNMENTAL ENTITIES.[75]

. . . .

Congress expanded the workload and the authority of the FCC with the Communications Satellite Act of 1962. This national policy established precedents which had to be considered in the DOMSAT policy-making process. This Act only altered the framework of the Communications Act of 1934 for the accommodation of satellites in 1962. The Commission's competitive perspective was still possible primarily through the flexible interpretation of its powers and functions under the Communications Act.

The role of Congress in the area of telecommunications policy was seen as an active one over the communications satellite issue, primarily because this

---

[74] 2 FCC 2d 671, Letter to ABC, FCC 66-206.

[75] 2 FCC 2d 668, Docket No. 16495, FCC 66-207.

new technology had the potential of far-reaching implications for the public interest. Congressional activity was also significant during the DOMSAT proceedings (as discussed in Chapter III) but then it was in the form of FCC oversight hearings rather than lawmaking debate. The primary argument over satellite communications concerned the question of ownership. Although there was much disagreement over alternatives, attitudes seemed to generally fear monopoly structures and favor competitive opportunities.

The ITU has played an important role in the field of international communications for more than a century and offers a way to prevent clashes between domestic, regional, and global systems. The satellite complicates this coordinating function because of its insensitivity to natural or political boundaries and because of the orbit usage considerations of communications satellites. Consequently, the need for the ITU's involvement in communications satellite activities is well recognized.

It was obvious then that some ongoing negotiating process involving the nations of the Americas is also a necessity in the management of the space spectrum for DOMSAT and regional systems. The orbital arc was definitely a new consideration in the process; but it was not a constraint. A single satellite could provide domestic coverage to the U.S. However, the greater the portion of service arc available, the greater the number of systems that could be configured for different purposes by different interests. The point is that radio frequency management, including orbit considerations, is a continuing process which is not amenable to exact legislative determination.[76] Leland Johnson of the Rand Corporation has recently noted that:

> One of the most interesting issues for the future will involve the interface between domestic and international service, including questions of expanding the number of international gateways, and the integration of domestic, regional and international satellite use, which may involve some restructuring of INTELSAT.[77]

Finally, the DOMSAT question was raised by the competitive spirit of industry, not by Congress, the Commission or the carriers. This forced the Commission to respond to the imperatives of this new technology, to the user demands and to the market structure with a viable policy. The domestic carriers dwarfed the international carriers and represented more powerful interests with more money at stake. However, the COMSAT proceedings had given the Commission no indication

---

[76]Smythe, "The 'Orbital Parking Slot' Syndrome and Radio Frequency Management," pp. 9-10.

[77]Letter from Mr. Leland L. Johnson, Director, Communications Policy Program, The Rand Corporation, to Robert S. Magnant, May 6, 1976. See Galloway, The Politics and Technology of Satellite Communications, Chapter 5 for an expanded discussion of the International Telecommunications Satellite Consortium (INTELSAT).

of what the domestic market's projected demands were; the amount of additional communications and new services that this technology could provide or that the market would support was still unknown. If a COMSAT-type monopoly [or COMSAT itself] was chosen as the structure for establishing domestic satellite services, market support would have been a less critical factor. However, regulated monopoly was not necessarily the ideal device for insuring a zealous, continuous quest for improved quality and variety of service.[78]

---

[78] Kahn, The Economics of Regulation, p. 141.

CHAPTER III

THE DEVELOPMENT OF THE DOMSAT POLICY

Policy is a definite course of action selected from among alternatives and in light of given conditions to guide and determine present and future decisions. However, when the process that generated that policy is reviewed in retrospect, it is important to look not only at the alternatives and the given conditions that were placed before the policy maker but also at the cross-section of sources from which these offerings came. In the DOMSAT considerations, market interrelationships, policical interrelationships and technological interrelationships were extremely important.

Since no one had a monopoly on the public interest, the Commission had to take into account all elements affecting the issue. Figure 2 is representative of the factors that the Commission repeatedly deals with. All played a part in the formulation of DOMSAT policy. The White House played an expanded role in this policy-making process and technology was seen as a force of increasing significance, promoting an even wider competitive environment and parallel Commission policies as

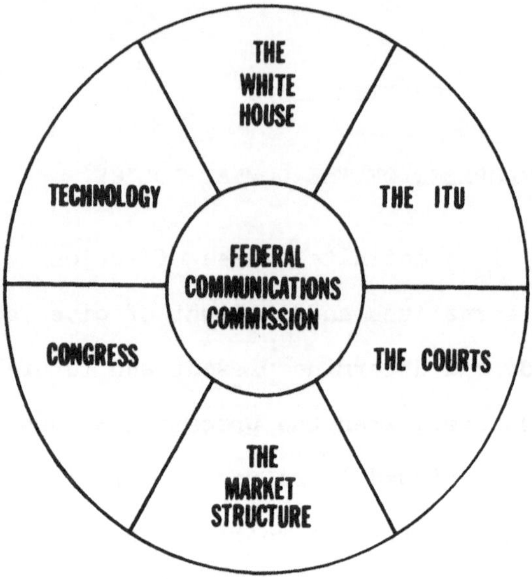

Figure 2. The World of the FCC

precedents of DOMSAT.

A variety of policy makers and personalities was very much in evidence throughout the DOMSAT proceedings. Administrations changed as did Commissioners during the almost seven year period; old voices provided comment and new voices raised new questions. It is difficult, if not impossible, to determine which of the forces were dominant in the DOMSAT decision but it is essential that the existence of all of them be recognized. However, one must bear in mind that only one policy-making body, the Commission, was responsible for balancing the interests of all concerned and making the final policy decision.

A. The Basic Issues

Not long ago the accepted standard for the economics of communications in the United States was regulated monopoly. In the changing social philosophy of the 1960's this standard held for local public telephone service but not for long-distance or private line services by ground microwave systems or satellites. The FCC decision on DOMSAT reflected this change and had an impact on the organization of a substantial part of the communications industry.[1]

The COMSAT Act left the question of who has the right to put up and operate a domestic satellite system shrouded in ambiguity. Administration and Congressional experts who drafted the Act assumed that satellites would be used mainly for international communications, at least over the next decade or so and failed to anticipate the speed with which satellite technology would develop.[2]

In the international area, the politicians had made the decision as to how communications satellite

---

[1] John McDonald, The Game of Business (Garden City, New York: Doubleday, 1975), p. 290.

[2] Charles E. Silberman, "The Little Bird that Casts a Big Shadow," Fortune, 75 (February, 1967), pp. 110-111. The assumption was that, for the immediate future, the only systems that could be operated would be low or medium altitude "random" satellites like Telstar which were basically not practical for domestic systems. However, Hughes Syncom II was operational in July 1963.

technology was to be adapted to the established institutions. The question was whether the existing structures of regulation and the domestic industry could be relied upon to implement this technology domestically or whether change was necessary to assure public benefit. Congress used its favorite device of saving most of the toughest problems for administrative resolution.[3] The effect of such innovation on the existing market was one of the more complex considerations addressed by the FCC.

Technology was compounding the issues in several ways. Not only had the Hughes geosynchronous satellite technology made equatorial orbital slots a desirable commodity internationally, complicating the definition of "domestic" systems, but now computer technology was both assisting in performing communications functions and increasing the demand for communications to provide distributed computing functions. Such demands raised new questions of competition simultaneously with DOMSAT, promoted new policies and altered the industry's shape a bit more.

1. A Notice of Inquiry

ABC's proposal for the construction and operation of a satellite for a private, specialized domestic

---

[3] Bernard Strassburg, "New Technology and Old Institutions", Telecommunications (June, 1974), p. 24.

communication requirement presented the Commission with significant questions as to the compatibility of such a proposal with the purposes, policies and objectives of the Communications Satellite Act. Believing that the public interest would best be served by obtaining the views of all interested parties before action was taken, the Commission in a Notice of Inquiry invited comments from all concerned on the following specific questions:

(a) Whether, as a matter of law, the Commission could promulgate policies and regulations, looking toward the authorization of nongovernmental entities to construct and operate communication satellite facilities for the purpose of meeting their private or specialized domestic communications requirements.

(b) Whether, as a matter of policy, it would be in the public interest to grant such authorizations and would it be technically feasible to accommodate the space services contemplated in light of existing electromagnetic interference criteria.

(c) Whether any such authorizations would impact upon the policies and goals set forth by the Communications Satellite Act and upon the obligations of the U.S. Government as a signatory in INTELSAT.

The Commission would also entertain other relevant matters to which the respondents wished to address themselves. This was the beginning. The issues of competition and ownership were being raised again and it would be almost seven years before a domestic satellite policy would be finalized.

a. The Ford Proposal

Of all the briefs, studies or statements filed with the Commission [19 parties responded on or before August 1, 1966], it was the Ford Foundation's proposal that brought the questions about a domestic satellite system to the forefront of public discussion and government decision.[4] In addition to answering the questions posed by the Notice of Inquiry, the Ford Foundation submitted a model of a private satellite system to provide for the transmission of both commercial and noncommercial television programming and called for the creation of a Broadcasters' Nonprofit Service Corporation (BNS) to establish such a system.[5]

The Ford plan was to provide a wholly new order of service--six channels for commercial networks and five noncommercial channels in each of the four time zones. This new satellite system, which had been developed by engineers, scientists and economists borrowed from Hughes Aircraft, IBM, the Rand Corporation and several universities, could provide greatly expanded

---

[4]Silberman, "The Little Bird", p. 111.

[5]FCC 2d 86 Report and Order, Appendix B at 108. This summary, which was provided with the First Report, provides excellent background on the initial responses. Appendix C discusses the legal considerations and Appendix D is the Technical Annex.

and improved transmission at much lower cost to the commercial television networks. Similarly ABC had estimated that it could reduce its own AT&T long-line costs by as much as 30 percent by switching over to satellites.[6] The system would also provide a "social dividend" in the form of free channels to link the Nation's now isolated educational television stations into one or more national networks, as well as free channels for instructional television for schools and colleges.[7]

In Congress, the Senate Subcommittee on Communications was holding hearings on progress in space communications and the Ford Foundation's proposed BNS service. Senator John O. Pastore, the Subcommittee's Chairman since 1955, had instructed his Communications Counsel Nicholas Zapple[8] to get in touch with the carriers, the networks, and anyone who might be concerned with these issues "so we might have the opinion of all. . . .[and] know from the very

---

[6]Andrew R. Horowitz and Bert Cowlan, "Should People Fight for Satellites," Tele-VISIONS (January/February, 1976), p. 5.

[7]Silberman, "The Little Bird", pp. 111, 223.

[8]At the time a veteran of more than twenty years of Congressional experience, Zapple had earned a reputation as a resourceful and innovative staff assistant. His knowledge of federal communications law was considered unsurpassed on Capital Hill. As noted in Peabody, Berry, Frasure and Goldman, To Enact a Law (New York: Praeger Press, 1972), p. 91.

beginning what the impact of these proposals will be. . ."[9] Reporting to this subcommittee on the FCC's progress in this area to date, Commission Chairman Rosel H. Hyde noted that comments had been filed by a number of parties on August 1, 1966 and that reply comments[10] were due for filing on or before October 1, 1966. When Pastore asked him, "As you sit here this morning, do you feel that you should make any further requests of the Congress in order to expedite the development of the communications area?", Hyde replied, "No, I do not, Mr. Chairman. I believe the direction that we have from Congress will permit us to proceed expeditiously."[11] Hyde did offer, however, this perspective of the difficulties that the Commission was having in the development of policy in the communications satellite area:

> While significant determinations have been timely made since February 1965 in each of these areas [the determination of the COMSAT system's technical description and the criteria for its ownership and operation], other problems of equal complexity and far-reaching effect have arisen

---

[9]U.S. Congress, Senate, Committee on Commerce, Subcommittee on Communications, Progress Report on Space Communications, Serial 89-78, August 10, 17, 18 and 23, 1966, 87th Congress, 2nd Session, p. 3.

[10]Under the Administrative Procedures Act, reply comments are routine for the presentation of opposing views and comments.

[11]U.S. Congress, Progress Report on Space Communications, pp. 17, 19-20.

> [DOMSAT]. This is, of course, to be expected in a field with a technology as fast developing and dynamic as communications via satellite.[12]

Pastore's boss, Senator Warren Magnuson (Chairman of the Commerce Committee) was heard later that year, while touring a COMSAT earth station, to say:

> I am hopeful that it [the Ford Foundation plan] won't interfere with the operation of COMSAT because we made a commitment to the American people when we passed the bill: that if they put their investment in COMSAT and gave it their support, then we would do what was best government-wise and regulation-wise to keep it going.[13]

If Congress had chosen to promote a total monopoly posture for COMSAT over all sky-circuitry, such would have been the determination of DOMSAT policy. Fortunately this was not the case.

After reviewing the August 1 comments of all interested parties, the Commission decided that it was necessary and "in the public interest" to expand the

---

[12]Ibid., p. 4.

[13]Michael E. Kinsley, Outer Space and Inner Sanctums: Government, Business and Satellite Communications (New York: John Wiley and Songs, 1976). Kinsley devotes approximately 20 pages (pp. 138-158) to the variety of supportive and opposing postures that the prime contestants took in their vying for control of DOMSAT. Although his conclusions are quick to categorize the FCC commissioners as "inferior minds or, at the very least, minds somewhat distracted from the public weal," his description of the machinations of industry gives the reader an appreciation for what the Commission was having to deal with.

initial scope of this proceeding and to invite comment on the Ford proposal. It additionally found that the initial filings were not fully responsive to the technical questions set forth in the initial notice. Accordingly, on October 20, 1966, the Commission adopted a "Supplemental Notice of Inquiry" for additional commentary. By December 16, 1966, 21 parties, nine of which had not filed previously, responded.[14]

Four parties [ABC, Ford, COMSAT and AT&T] had submitted proposals for domestic satellite systems. Those of COMSAT and AT&T were counter-proposals to the ABC and Ford proposals aimed at retaining the former's pre-eminence in domestic and space communications activity.[15] They urged the FCC to authorize general purpose systems while the latter proposals asked for special purpose systems for the distribution of television programs, anticipating cost savings made possible by satellite technology. During the course of the proceeding both COMSAT and Ford supplemented their original proposal by suggesting pilot programs rather than full-scale operational domestic systems. The others who made their views known to the Commission included common carrier, broadcast, computer, aviation,

---

[14] 22 FCC 2d 111.

[15] Herbert I. Schiller, "Communications Satellites: A New Institutional Setting," Bulletin of the Atomic Scientists (April, 1967), p. 8.

press, government and educational interests. In addition to the ownership problem, a wide disagreement on a broad range of legal, technical and economic issues emerged.[16]

COMSAT's pilot proposal contemplated an investment of $58 million in the space and ground segments to place a limited satellite system in service by 1970. This pilot program was to last only two years (1970 and 1971) and COMSAT would own and control the system only as a trustee or steward, with final arrangements as to ownership and control to be postponed until the completion of the program. AT&T expressed support for the pilot program but asked that it, rather than COMSAT, be permitted to own the ground stations.[17]

In 1968, Chairman Hyde summarized this early activity as follows:

> With respect to domestic satellites, it is reasonable to look forward to the time when satellites will be

---

[16]Final Report, President's Task Force on Communications Policy, December 7, 1968, (Washington, D.C.: U.S. Government Printing Office, 1969), 0-351-636, Chapter 5, p. 2. A fair amount of disagreement among owners, operators and regulators of satellite systems still exists today. For this reason, Hughes Aircraft (a principal builder of satellite systems for many parties) has established a firm policy not to participate in any evaluation of regulatory actions. Letter from Dr. Albert D. Wheelon, Vice President and Group Executive, Hughes Aircraft Company to Robert S. Magnant, May 4, 1976.

[17]Henry Geller, "Competition and Monopoly Policies in Domestic Satellite Communications," The Antitrust Bulletin, 13 (Fall, 1968), pp. 954-955. Geller was FCC General Counsel at the time.

> competing with, or complementing, terrestrial microwave and cable for long-distance telephone and other point-to-point communications services. The issue of how to structure the ownership, operation and use of a domestic satellite system is now before us in a general public inquiry. Its purpose is to resolve a number of difficult legal, technical and policy questions involved in the use of satellites by both carrier and noncarrier entities. We have under consideration a pilot program to assist in accumulating data toward formulation of final format for our domestic system.[18]

b. A Question of Law

In considering the question of competition or monopoly as regards DOMSAT systems, it was first necessary to look at the threshold legal questions, which could have severely limited the Commission's options with respect to ownership and control of a domestic satellite system.[19] The primary question, of interest to all concerned, was whether the Communications Act of 1934 or the Communications Satellite Act of 1962 would regulate domestic satellite communications. Another concern raised was whether the authorization of DOMSAT systems would violate our international agreements of 1964 as

---

[18]Rosel H. Hyde, "The Role of Competition and Monopoly in the Communications Industries," The Antitrust Bulletin, 13 (Fall, 1968), p. 908.

[19]Geller, "Competition and Monopoly Policies in Domestic Satellite Communications," p. 955.

members of INTELSAT. The various interests involved tended to interpret the applicability of these laws and agreements according to their own interests.

COMSAT naturally believed that under the 1962 Act it was the only entity authorized to have satellites, be they domestic or international, and the Ford Foundation believed that under the 1934 Act the FCC had the legal right to authorize an entity like BNS. Although some carriers asserted that the Commission had the broad powers to authorize nongovernmental entities to establish and operate DOMSAT facilities, others alleged that the legal capacity was lacking. However, all carriers agreed that such systems should not be authorized to noncommon carriers as a matter of policy. The Commission concluded that both Acts clearly empowered it to authorize DOMSAT facilities to be owned by any entity as the national public interest required and that INTELSAT agreements would not be compromised. Pertinent arguments supporting these conclusions follow.[20]

(1) The Communications Act of 1934. The broad language of this act had provided the Commission with "regulatory power over all forms of electrical

---

[20] 22 FCC 2d 128, Appendix A, Memorandum on Legal Issues. Much of the following information is more extensively developed in the Ford Foundation's legal brief; prepared in response to Docket No. 16495 by Ginsburg and Feldman, Attorneys for the Foundation, dated August 1, 1966.

communications. . . ."[21] Section 303(g) of the Act specifically directed the Commission to "generally encourage the larger and more effective use of radio in the public interest." In 1961, the Commission concluded that it had the power to authorize the private use of communications satellites and responses from its inquiry on the subject supported its position.[22] Prior Judicial and Commission decisions also supported the position that Congress had given the Commission broad power. In NBC v. United States, for example, Justice Frankfurter stated that "Congress endowed the Communications Commission with comprehensive powers to promote and realize the vast potential of radio."[23]

(2) The Communications Satellite Act of 1962. This Act did not preclude Commission authorization of private noncommon carriers to construct and operate DOMSAT facilities. The primary purpose of the Act was to establish an international communications satellite system and was intended as a major step towards a

---

[21]U.S. Congress, Senate, Committee on Interstate Commerce, S. Rep. No. 781, 73rd Congress, 2nd Session, April 17, 1934, p. 1. See also Chapter I, note 22 and Krasnow and Longley, note 1, supra, p. 15.

[22]See Docket No. 14024, Inquiry Into the Administrative and Regulatory Problems Relating to the Authorization of Commercially Operable Space Communications Systems, March 29, 1961, reprinted in U.S. Congress, House, Committee on Science and Astronautics, Hearings on Communication Satellites, May 8, 9, 10 and July 13, 1961, 87th Congress, 1st Session, pp. 537-581.

[23]319 US 190, 217 (1943). See also Chapter I, notes 25 and 26.

"global communications network."[24] An international system was seen both as a means to move ahead of the Soviet Union (which had apparently taken the lead in space technology) and as a way for the United States to assert her leadership at the 1963 Extraordinary Administrative Radio Conference of the International Telecommunications Union (called in 1963 to allocate frequencies for communications satellite systems).[25] Since Telstar was seen as the COMSAT model for many years to come, its random orbit made sense only in the context of an international system.[26] The language of Section 102(d) of the Act expressly recognized the power of the Commission to authorize private DOMSAT systems. Mr. Dingman, in written testimony to the Senate on another occasion, had noted that "sometime in the future it may be necessary to augment the initial system and this could involve the establishment of a new satellite system."[27]

(3) <u>The 1964 International Communications-Satellite Agreements</u>. These agreements did not

[24] See Chapter II, note 27.

[25] U.S. Congress, House, Committee on Government Operations, H.R. Rept. No. 178, 89th Congress, 1st Session, March 17, 1965, p. 22.

[26] See Chapter II, notes 23 and 30.

[27] U.S. Congress, Senate, Hearings on Communications Satellite Legislation Before the Senate Committee on Commerce, April 10-13, 16, 24 and 26, 1962, 87th Congress, 2nd Session, p. 202. This was in response to questions that had been raised by Senator Yarborough.

preclude the establishment of DOMSAT facilities by noncommon carriers. The Inter-Governmental Agreement and the Special Agreement (concluded in 1964) were designed to establish "a global commercial communications satellite system." Nothing in the language of either instrument has been interpreted as precluding separate domestic systems.[28] Like the 1962 Act, they were aimed at creating a single system to serve the international needs of the more than 50 signatory nations. It was never intended to serve the membership's domestic requirements or to prohibit them from developing separate systems to meet those requirements. A possible conflict of interest could have developed if COMSAT, as the international "system manager," had undertaken a separate system.

2. Technology

The merging of the computer and communications technologies in the late 1960's only complicated the Commission's problem in developing a DOMSAT policy. As the need for computers to talk to each other increased, the wideband communications capability of the satellite became more attractive. DOMSAT would definitely increase the options available for

---

[28] U.S. Department of State, Treaties and Other International Acts Series No. 5646, International Telecommunications Satellite Consortium (INTELSAT), 1964.

the computer communications requirements.[29] As Dr. Norman Abramson of the University of Hawaii recently noted:

> . . .long-haul service provided by satellite is usually more cost-effective than comparable services provided by purely terrestrial links. This is especially true of specialized, higher rate services. Today there is little, if any, terrestrial capacity available to provide long-haul transmission of data at rates of one million bits per second and higher. To provide new services at these rates in most situations will require both large inflexible investments and new construction of terrestrial facilities, which will require long implementation times. Satellite systems can be quickly and inexpensively installed to establish circuit connections wherever the user desires and at the data rates that he needs.[30]

The DOMSAT issue thus had at its base a persistent driving technology--a technology that had blurred a once comfortable distinction between communication and data processing.[31] Not only were the two technologies

---

[29]Letter from Dr. Robert W. Lucky, Head, Transmission Terminal Systems, Bell Laboratories, to Robert S. Magnant, May 22, 1976.

[30]Norman Abramson and Eugene R. Cacciamani, Jr., "Satellites: Not Just a Big Cable in the Sky," IEEE Spectrum (September, 1975), p. 36. SBS plans to offer such a service, tailored to each customers requirements, as part of its integrated (voice, image, data of different speeds) offering.

[31]Bernard Strassburg, "Communications and Computers: How Shall the Twain Meet?", Public Utilities Fortnightly, 82 (September 12, 1968), p. 74.

becoming more interdependent, but they were doing it at phenomenal speed. The Commission became concerned over what the resulting impact on communications might be and initiated a Notice of Inquiry [hereafter referred to as the Computer Inquiry] to determine policy on the subject. This Inquiry was in progress throughout practically the entire DOMSAT proceeding.

To facilitate an understanding of this marriage between computers and communications, definitions for such terms as message switching systems, time-sharing and distributed networks are developed below, along with some simple descriptions of the technologies involved and some perspectives on the implications of "computercations" and the Computer Inquiry on the Commission and DOMSAT.

a. Technology Dynamics

Abramson sees the force of technology and technological change on the communications industry as probably one of the most significant factors that has influenced the DOMSAT policy.[32] Because of the extraordinary advances in satellite technology and because the Federal government itself became involved in competitive enterprise through the creation of COMSAT, no more cogent example of the force of the

---

[32]Letter from Dr. Norman Abramson, Director, the ALOHA System to Robert S. Magnant, May 4, 1976.

technological change upon economic systems can be found than the study of the evolution of Federal policies with respect to the burgeoning telecommunications industry and DOMSAT in particular.[33]

It should be noted that the technology that spawned the common carrier communications industry was one of analog transmission, frequency division multiplexing, and electro-mechanical switching. With the emergence of computer time-sharing technology in the mid-1960's, digital transmission, the continuing increase in computer applications, and new requirements for "machine communications", the idea of creating new and different transmission systems to meet these requirements was quite logical. This was evident even to the established carriers. As the Bell Telephone people put it, in a 1957 technical paper on data transmission services:

> The telephone network was developed for speech transmission and its characteristics were designed to fulfill that objective. Hence, it is that the use of it for a distinctly different purpose such as data transmission may impose compromises both in the medium and in the special service contemplated.[34]

---

[33] Thomas P. Murphy, "Federal Regulatory Policy and Communications Satellites: Investing the Social Dividend," The American Journal of Economics and Sociology, 31 (October, 1972), p. 338.

[34] John D. Dingell, "The Role of Spectrum Allocation in Monopoly or Competition in Communications," Antitrust Bulletin, 13 (1968), p. 941.

Obviously, there would be requirements for systems for data communication services where the medium was not a limitation and compromises would not be imposed.

Within a period of only two decades, the electronic computer advanced from an esoteric tool of the scientist to a pervasive participant in the processes of government, business, and education. In this short time, the computer matured from a laboratory curiosity to the world's fastest growing industry.[35] As late as 1959, however, electronic data processing was generally limited to in-house or intrafirm/plant concepts.[36]

The computer itself is a giant communications system with complex switching techniques. As the power and utilization of computers increased, it was only natural to connect the internal communications circuits of the computer to existing external communications circuits and thus provide wide geographic dispersion to the capabilities of a computer.[37]

From the communications industry perspective,

---

[35] Barry Taub, "Federal Communications Commission Regulation of Domestic Computer Communications: A Competitive Reform," Buffalo Law Review, 22 (Spring, 1973), p. 947.

[36] Strassburg, "Communications and Computers: How Shall the Twain Meet?", p. 69.

[37] Ervin K. Dorff, "Computers and Communications: Complementing Technologies," Computers and Automation (May, 1969), p. 22.

the sheer growth in message volume increased the number of switching networks needed and the complexity of the sophisticated control mechanisms required. The computer represented a tool that was naturally designed to easily perform this switching function. It is in this manner that the two technologies complement and support one another--communications facilities are used to expand and extend the capabilities of computers; computers are used to control and expand communications systems.[38]

There are also strong economic and social influences promoting increased cooperation between telecommunications and computer interests. It is important to recognize that this cooperation represents far more than a scattering of technical improvements. There has been a fundamental change in the scope and utility of all our resources for gathering, exchanging and using information.[39] Schiller has labeled today's telecommunications technology as one of the most dynamic forces, affecting not only the ideological but the material bases of society.[40] With the broad

---

[38] Ibid.

[39] Robert L. Werner, "A Lawyer Looks at Our Communications Policy," Jurimetrics Journal, 11 (December, 1970), p. 81.

[40] Herbert I. Schiller, Mass Communications and American Empire (New York: Augustus M. Kelly Publishers, 1971), p. 33.

implementation of "distributed computer" concepts, the usefulness of interactive, automated information processing systems can be further extended to the general populace on a shared basis.[41]

b. Computercations

The term "computercations" has been proffered here not just to keep the etymologist of tomorrow employed but rather to specifically indicate an intimate relationship between the computer and communications disciplines. This is not meant to imply simply the tying of a telephone to a computer, but rather the distribution of computation functions throughout a medium of communications, to any or all terminal points.[42]

Computers became linked with communications services at the Massachusetts Institute of Technology with the inauguration of Project MAC in 1961, which is believed to be the initial mating of communications and computers. It was now possible for the user to reach the computer from remote locations via communications channels furnished by common carriers and to thereby have on-line and real-time access to the

---

[41]Taub, "FCC Regulation of Domestic Computer Communications: A Competitive Reform," p. 949.

[42]A "distributed" computing system can either take the form of remote input devices interconnected to a central processor, or the addition of intelligent terminals, or interconnected computers.

computer as well as instantaneous response from it.[43]

Several major technological developments have fostered this increasingly close relationship between data processing and communications. One major breakthrough was the development of input-output devices, called terminals, that could be used to transmit and receive data (to and from a computer) at a remote location via the common carriers' facilities. However, an initial shortcoming of computer systems was that they were capable of processing only one job at a time. Time-sharing technology, where new hardware and software enabled third-generation computers to process several jobs simultaneously, significantly altered this situation.[44]

Digital transmission is undoubtedly one of the more significant developments effecting the conversation of machines. Computers "talk" in digital mode--a series of "on" and "off" pulses. Most conventional communications systems transmit in analog format--a continuous wave-like signal. Noise and interference, as well as signal, are amplified in the process of transmission. A listener can screen out such unwanted noise but a machine cannot. For computercations,

---

[43] Strassburg, "Communications and Computers: How Shall the Twain Meet?", p. 70.

[44] Notes, "The FCC Computer Inquiry: Interfaces of Competitive and Regulated Markets," Michigan Law Review, 71 (November, 1972), pp. 173-174.

transmission accuracy becomes much more critical. Digital transmission provides just such performance and the requirement of modems, which are translation devices that convert digital outputs to analog formats for transmission over conventional networks, is eliminated and along with it a great deal of expense and a potential source of error.[45] Another benefit of a digital system is reduced costs. Digital technology can squeeze more messages into each circuit, thus reducing costs by improving transmission efficiency.

The combination of rapidly advancing technology and vigorous competition in the computer industry has decreased the cost of computer time. Consequently, as computation costs have fallen, communications costs have become more and more significant.[46] Lawrence Roberts of Telenet Communications and formerly with the Office of the Secretary of Defense's Advanced Research Projects Agency estimates that the crossover point, where the cost of using computers to allocate bandwidth [time sharing the frequencies by means of computer switching] became cheaper than the cost of the communications themselves, occurred during 1969.[47]

---

[45] Wynn Walters, "Computer Communications: The Start of a Revolution," The Business Quarterly (Winter, 1973), p. 83.

[46] Notes, "The FCC Computer Inquiry," p. 175.

[47] Lawrence G. Roberts, "Data by the Packet," IEEE Spectrum (February, 1974), pp. 46-51.

Bernard Strassburg, Chief of the FCC's Common Carrier Bureau at the time, suggested that the common carrier communications industry was "functioning in the eye of a hurricane surrounded by unprecedented turbulence."[48] In 1900, there were few who could foresee the impact and implications of the Gibbs concept of describing the universe in terms of probabilities. Similarly today, it is equally unlikely that more than a handful of people fully perceive the full impact of the combining computers with telecommunications.[49]

We may speculate ad infinitum as to the possible applications to which computer technology may be put in the future. However, one thing is certain; a communications link will provide the means by which its potential will be fully realized.[50] In providing such links, the satellite's capabilities far surpassed

---

[48] Horace P. Moulton, "Monopoly and Competition Issues Facing the Communications Industries," The Antitrust Bulletin, 13 (Fall, 1968), p. 889.

[49] Stanley Winkler, "Computer Communication--the Quiet Revolution," Computer Communication--Impacts and Implications, Stanley Winkler, editor, ACM/ICCC, 1972, p. 29. J. W. Gibbs (1839-1903) was a mathematical physicist who wrote of, among other things, the electromagnetic theory of light. His most important publication was his famous paper, "On the Equilibrium of Heterogeneous Substances." Only after engineering and physics had considered the science of distributions were the developments of modern communications possible.

[50] Strassburg, "Communications and Computers: How Shall the Twain Meet?", p. 70.

those of the existing terrestrial systems.

Business demands computercations. With their enormous appetites for information, computers can collect, assemble and analyze vast amounts of information and give managers the data they need. Telecommunications gets the information where it is needed, when it is needed and in the form it is needed. As computers and telecommunications become even more inseparable in the future, the delineation between the processing function, the switching function, and the transmission function will become increasingly blurred. In a digital network, computers will take on the appearance of "fat spots" of data in the transmission links.[51] In an FCC public inquiry on computers and communications, initiated in 1966 and extending to 1971, specific attention was focused on these growing interrelationships.

c. The Computer Inquiry

Within the same year that the question of domestic satellite communications was raised, the basic definition of communications was also being questioned as a result of technological trends. On November 10, 1966 the Commission launched the Computer Inquiry to explore the broad range of regulatory and policy problems generated by these technological

---

[51] Walters, "Computer Communications," p. 81.

developments.[52] While some observers believed that the giant strides made in the technologies of computer components and telecommunications would inevitably lead to a "total computer telecommunication" utility,[53] the threshold question of the inquiry was namely, where was the "line of demarcation" between communications and data processing?[54] The questions posed by the Inquiry and the subsequent responses and replies ranged from consideration of the current and future trends relating to the computer and information processing industry; to the adequacy of existing legislation to deal with such trends; to the need for new measures "to protect the privacy and proprietary nature of data. . .transmitted over communication facilities. . ."[55]

Over 3000 pages of material were submitted from the communications industry, the data processing industry, common carriers, users, government and

---

[52] 7 FCC 2d 11, Docket No. 16979, Regulatory and Policy Problems Presented by the Interdependence of Computer and Communication Services and Facilities, 1966.

[53] Robert A. Dunlop, "The Emerging Technology of Information Utilities," The Information Utility and Social Change (Montvale, New Jersey: AFIPS Press, 1970), p. 45.

[54] Strassburg, "Communications and Computers: How Shall the Twain Meet?", p. 73.

[55] Robert E. Lee, "The Role of the Federal Communications Commission," Computer Communications--Impacts and Implications, ACM/ICCC, 1972, p. 49.

other interested parties. An explanation of the categories of systems that were discussed is necessary for an understanding of the growing computer/communication relationships:

> (1) Data processing--the use of a computer for the processing of information as distinguished from circuit or message switching. Processing involves the use of the computer for functions such as storing, retrieving, sorting, merging and calculating data according to programmed instructions.
>
> (2) Message switching--the computer-controlled transmission of messages between two or more points via communications facilities wherein the content of the message remains unaltered. The Commission has specifically excluded this from the data processing category.
>
> (3) Hybrid data processing--a service offering wherein the message-switching capability is incidental to the data processing function of purpose.
>
> (4) Hybrid communications--a service offering wherein the data processing capability is incidental to the message-switching function or purpose.[56]

Although the continual development of the computercations technology makes classification difficult and while the distinction between message switching and data processing loses validity in direct proportion to the sophistication and complexity of the computer hardware and software, these categories are the best definitions currently available for this discussion.[57]

---

[56]Notes, "The FCC Computer Inquiry," pp. 178-179.

[57]Delbert D. Smith, "The Interdependence of Computer and Communications Services and Facilities:

The FCC's primary interest was with the communications common carriers; theirs was a regulated market which operated under tariffs, schedules of rates and charges. The dynamic computer industry represented the competitive market; driven by innovation and free enterprise. The line between hybrid communications and hybrid data processing thus was important because it marked the boundary between tariffed and nontariffed services.[58] The distinction between hybrid services was subtle and it became more complex with each technical innovation. The most obvious form of hybrid communications function performed by a computer is store-and-forward message switching, in which the computer acts as an intermediary between the sender and the receiver.[59]

There were three groups of competitors [and still are] that could influence the future of the data processing field--the computer industry, the service bureau industry (which performs data processing

---

A Question of Federal Regulation," University of Pennsylvania Law Review, 117 (April, 1969), p. 831.

[58]The Bell System is precluded from offering nontariffed services by its 1956 consent decree. In addition, Bell spokesmen have consistently disavowed any intent to compete in the field of computer manufacture or data processing service. See Strassburg, "Competition and Monopoly in the Computer and Data Transmission Industries," The Antitrust Bulletin, 13 (Fall, 1968), p. 993.

[59]Notes, "The FCC Computer Inquiry," p. 183.

on a fee or contract basis and includes specialized subscription services) and the communications common carriers. As the data processing industry grew, it approached "unprecedented intimacy" with the communications industry.[60] What provided specific appeal to all was the market's potential. The number of data terminals installed in the United States during 1970 was about 185,000 and was estimated to reach 2,500,000 by 1980. It was also anticipated that the number of data transmissions per year would jump from almost 15 billion to 248 billion by the decade's end.[61] Today, data transmission services comprise less than two percent of the entire loading of any system (or approximately two percent of the telephone traffic) but are estimated to be growing at the rate of fifty percent per year. It has been estimated that by 1984 as much as ninety percent of the computers used in the United States will be tied into the public communications system.[62]

In their Inquiry comments to the Commission

---

[60] Notes, "The FCC Computer Inquiry," pp. 176, 181, 187.

[61] Taub, "FCC Regulation of Domestic Computer Communications," p. 950. See Nolan, "Moving Business Data is Big Business," Rutgers Journal of Computers and Law, 1 (Fall, 1970), pp. 2, 4.

[62] Ibid., p. 949; see also Computers and Telecommunications, Organization for Economic Cooperation and Development Information Studies, 3 (Paris: OECD, 1973), p. 11. Also noted again by R. E. Butler, note 63, infra, p. 50.

concerning the adequacy of the then available data transmission services, by and large some sixty respondents--with the exception of the common carriers--were critical of the service offerings then being provided. One of the larger problems that faced both the computer and communications industries was the lack of advanced communications capabilities in computer systems [to a large extent this remains a significant problem today].[63] The computercation requirements demanded digital transmission facilities with greater efficiencies. Quality, speed, capacity and economy, the kind of improvements that a satellite system could provide, were noted to be lacking.

IBM, in its response, commented:

> Data communications requirements have been met to date by adaptation of facilities designed primarily for voice services. While there are important advantages in using a single network, there are fundamental differences in the requirements of voice and data communications.

Bunker-Ramo Corporation observed:

> The existing costs of communications channels and facilities within the present tariff structure are an inhibiting factor to the widespread use of computer services.

Sanders Associates, Inc., a defense contractor and manufacturer believed that:

---

[63] R. E. Butler, "International Cooperation and Regulation--Foundations for Development," <u>Telecommunications</u>, 10 (January, 1976), p. 53.

> . . .our existing colossally complex telephone network, while manifestly indispensable for voice transmission, cannot economically be transformed into an efficient data transmission medium.

Martin Marietta Corporation, a diversified aerospace manufacturer, noted:

> . . .requirements of data communications must ultimately. . .be accommodated by specialized all-digital systems which. . .shall: (1) offer improved quality of data transmission, (2) provide faster access time, (3) provide higher bit rates [data rates]. . .

After considering such comments and various market studies, the FCC and its staff concluded that there was, in fact, public need for the proposed services and facilities, and noted that there was a large potential market yet to be developed.[64]

The Commission's Final Decision on the Computer Inquiry included the statement that: ". . .we are not proposing, at this time, to regulate data processing, as such, ..." However, because of the growing interdependence between the computer and the communications industry and the need to insure appropriate regulatory treatment, the Commission did recognize that

---

[64]Stuart L. Mathison and Philip M. Walker, "Regulatory Policy and Future Data Transmission Services," Computer Communications Networks (Englewood Cliffs, New Jersey: Prentice-Hall, Inc., 1973), pp. 342-344. Mathison and Walker also provide excellent coverage of the Computer Inquiry issues in Computers and Telecommunications, Chapter I, note 14.

there was a definite need for a maximum separation of activities which are subject to regulation from nonregulated activities involving data processing.[65] Although the Commission saw "no need to assert regulatory authority over data processing activities whether or not such services employ communications facilities in order to link the terminals of subscribers to centralized computers," it did retain the prerogative to "reexamine the policies set forth. . .if there should develop significant changes in the structure of the data processing industry."[66]

The Final Decision had not answered the question raised by the Justice Department regarding the Commission's jurisdictional authority. In its response to the Inquiry, Justice had concluded:

> It is our opinion that remote-access data processing is not common carrier communications and hence is not subject to the Commission's jurisdiction under Title II of the Communications Act.

The Commission, in concluding that regulation was unwarranted, had bypassed for the moment the question of whether or not it had the necessary legal authority to impose regulatory controls upon these services.

---

[65]Lee, "The Role of the Federal Communications Commission," p. 49.

[66]Stuart L. Mathison and Philip Walker, "Regulatory and Economic Issues of Computer Communications," Proceedings of the IEEE, 60 (November, 1972), p. 1257.

But in its decision it did conclude that it had "ample jurisdiction" to specify conditions under which common carriers may provide data processing services. The "line" became increasingly more difficult to define.[67] In the "hybrid service" area the Commission adopted a "wait-and-see" attitude, deciding that it did not have the Solomon-like wisdom needed to separate hybrid communications from hybrid data processing, and planned to treat such services on a case-by-case basis [in order that it might see how such systems would develop].[68] The Commission's decision, to regulate hybrid communications on a discretionary basis, as a practical matter, precluded data processing firms from offering such service because of the threat of regulation.[69]

One lesson that strongly emerged from the Inquiry was that the Commission henceforth would be concerned with the interdependence of two areas affecting our society, computers and communications.[70] Also decreasing computation costs, in conjunction with the fact that

---

[67]Ibid. Marcus Cohn, a former FCC lawyer, reviewed the Inquiry (see Rutgers Journal of Computers and Law, 1 (Fall, 1970)) in an article titled, "The Federal Communications Commission," to make the
Computer
point.

[68]Lee, "The Role of the Federal Communications Commission," p. 50.

[69]Notes, "The FCC Computer Inquiry," p. 182.

[70]Lee, "The Role of the Federal Communications Commission," p. 50.

voice telephone lines have only a limited capacity to handle data, would make any superior alternative communications system, like a domestic satellite, a welcome addition in the computer world.[71] But while the computer-communications explosion was real and accelerating, its full potential depended greatly on communications developments like the satellite which, while clearly visible, had yet to be broadly accomplished.[72]

3. The "New Competition"

When the Commission issued its Above 890 decision in 1959, it was by no means prepared to authorize any additional competition in the common-carrier communication business and made it clear that the door was only open to individual users who wish to meet their own individual needs.[73] However, the new wave of regulatory activity, which started in 1968 and opened up areas of common carrier service to competition, had a substantial impact on the formulation of the DOMSAT policy.[74] It put to the test of contemporary

---

[71] Taub, "FCC Regulation of Domestic Computer Communications," p. 960.

[72] Dorff, "Computers and Communications," p. 22.

[73] Alfred E. Kahn, The Economics of Regulation: Principles and Institutions, Vol. II (New York: John Wiley and Sons, 1971), p. 131.

[74] Letter from Mr. Andrew Margeson, Staff Economist, House of Representatives, Committee on Interstate and Foreign Commerce, to Robert S. Magnant, April 29, 1976. Dr. Manley Irwin, a former FCC

reasonableness the long-standing policies and practices of both the carriers and the Commission. It altered long-established institutional structures and conventions and opened the market place to the innovators; not only the innovators of electronic equipment but also the innovators of regulatory policy. Mr. Fred W. Henck, the editor of TELECOMMUNICATIONS REPORTS who has watched the industry's evolution for the past 35 years, sees the DOMSAT policy as a direct outgrowth of the Commission's interconnection and specialized carrier policies, both of which were developed while DOMSAT was pending.[75] This is also supported by Barry Taub in a 1973 article who referred to Carterfone (the interconnect policy) as the beginning of a second wave of competition which culminated in DOMSAT, "where the seeds planted by Above 890 began to flower."[76] Summaries of the interconnect and specialized carrier issues follow. These rulings made the integration of satellite systems with terrestrial networks possible.

---

economist, has identified DOMSAT as part of the Commission's new pro-competitive policy trend that was embarked upon in 1968. See Letter from Professor Manley R. Irwin, University of New Hampshire, to Robert S. Magnant, May 13, 1976.

[75] Letter from Mr. Fred W. Henck, Editor, TELECOMMUNICATIONS REPORTS, to Robert S. Magnant, May 17, 1976.

[76] Taub, "Federal Communications Commission Regulation of Domestic Computer Communications," pp. 966-970.

a. Interconnection

The Communications Act of 1934 specifically provided for the protection of the nation's telecommunications common carriers from outside competition through regulation as was noted in Chapter I. But in the strict sense this was not meant to be an eternal umbrella for the common carriers. A changing of the environment began almost twenty years ago; thoughts of private systems, foreign attachments and specialized services gave birth to the interconnection issue.

In recent years the question of interconnect, which includes not only the interconnection of subscriber-provided terminal devices, but also of independent communications systems (such as a domestic satellite system) to the telephone network, has generated considerable controversy. Historically, the telephone companies prohibited such interconnection by their subscribers, maintaining that it might jeopardize the telephone network if uncontrolled, but also recognizing that it would most certainly cut into the sales of their manufacturing subsidiaries (such as AT&T's Western Electric and GT&E's Automatic Electric).[77]

The carrier's arguments of potential system harm were generally accepted without question by the and the state public-utility commissions in the past

---

[77]Mathison and Walker, "Regulatory and Economic Issues of Computer Communications," p. 1260.

and consequently a seemingly impregnable wall was built around the telephone networks. However, the FCC's 1968 Carterfone decision caused the wall to crumble. Although Carterfone caused the most dramatic changes in interconnection regulation, it wasn't its beginning. Over a decade earlier the common carriers' blanket prohibition against interconnection of customer-owned equipment was tested before the U.S. Court of Appeals, in Hush-a-Phone Corporation v. United States.[78] This case concerned a rubber cuplike device designed to be attached to the microphone portion of the telephone handset to provide privacy in conversation; its use on the Nation's telephone network had been barred by provisions in the carrier's tariffs. Reviewing the Bell interstate toll tariff, the court found the ban to be illegal and ruled that it was ". . .an unwarranted interference with the telephone subscriber's right to use his telephone

---

[78]99 U.S. App. D.C. 190, 238 F 2d 266, D.C. Cir. (1956). Actually the question of interconnect was raised almost 30 years ago! See 11 FCC 1033, Docket No. 6787, In the Matter of Use of Recording Devices in Connection with Telephone Service. Telephone recording devices have been in use to some extent since 1916 (Edison, Inc. developed an acoustically coupled device in 1915 called the Telescribe). In its March 24, 1947 ruling the Commission found:

> insofar as any tariff regulations on file with us have the effect of barring such use of recording devices, such tariff regulations are unjust and unreasonable and therefore unlawful under the provision of section 201 of [the] Communications Act.

in ways which are privately beneficial without being publicly detrimental."[79]

AT&T was ordered by the court to revise its tariff to permit use of the Hush-a-Phone device. It did as ordered but the revised tariff retained the general interconnection prohibition. Consequently, in 1966, Carter Electronics Corporation brought an antitrust suit against the Bell System and GT&E of the Southwest.[80] The case was referred to the FCC. The Carterfone was an electronic device used for acoustically coupling the base station of a mobile radio system (or other private communication system) with the telephone network.[81] The telephone companies argued before the FCC that use of the Carterfone violated the integrity of the telephone system, which required the use of carrier-supplied attachments only. The Commission was not persuaded by this argument and in a unanimous opinion, in June 1969, found the tariff restrictions to be unreasonable, unlawful and discriminatory under the Communications Act of 1934.

The Commission further concluded:

---

[79]Ibid.

[80]Mathison and Walker, "Regulatory and Economic Issues of Computer Communication," p. 1261.

[81]Approximately 3500 of these devices were sold between 1959 and 1966. The telephone companies warned Carter's customers that the tariffs prohibited such devices and those who violated them risked having their telephone service terminated.

> . . .a customer desiring to use an interconnecting device to improve the utility to him of both the telephone system and a private radio system should be able to do so, so long as the interconnection does not adversely affect the telephone company's operations or the telephone system's utility for others. A tariff which prevents this is unreasonable; it is also unduly discriminatory where, as here, the telephone company's own interconnecting equipment is approved for use. The vice of the present tariff. . .is that it prohibits the use of harmless, as well as harmful devices.[82]

The Commission appropriately struck the unlawful tariff and permitted the carriers to propose new tariff provisions in accordance with its opinion and to specify technical standard for the protection of the telephone system against harmful devices if they so desired.

b. The Specialized Carriers

The MCI and Specialized Carrier decisions were extensions of the Above 890 decision. After the latter, if you had a requirement for your own private communications but you could not afford your own private system, what the common carrier was offering for services was all you could get. In 1963, Microwave Communications, Inc. (MCI) recognized this fact and petitioned the FCC for a license to offer "specialized

---

[82] Use of the Carterfone Device in Message Toll Telephone Service, 13 FCC 2d 420 (1968), pp. 424-425. Additional perspective on the impact of Carterfone is found in Mathison and Walker, note 66, supra, pp. 1262-1264.

services." It took six years for MCI's construction permits to be approved but in 1969 the precedent was set for the establishment of new carriers, primarily because the established carriers [as previously noted in the responses of the Computer Inquiry] were not providing the user with the services that were wanted.[83]

MCI went on the air as a full-fledged specialized carrier on January 1, 1972. If that were all that had happened, there would be little cause for more than academic interest in the "computercations" community. However, the FCC's 1969 approval of MCI's permits triggered over 1900 new microwave station applications from several dozen firms who were proposing to build new specialized-carrier facilities. One of the applicants, the Data Transmission Company (Datran) proposed to build a nation-wide, all-digital, switched network and offer exclusively data-transmission services on both a switched and private-line basis.[84]

These numerous and varied applications presented the FCC with a major policy problem, for the Commission was not sure that its MCI precedent should be extended on a nationwide basis without further analysis.

---

[83]Mathison and Walker, "Regulatory and Economic Issues of Computer Communication," pp. 1264-1266. A variety of dedicated private-line services (data services, or a combination of data and voice) in truth was more than the existing carriers were willing to provide or, in some cases, capable of providing.

[84]Ibid., pp. 1264-1265.

Therefore, in July 1970, it issued a public inquiry into the merits of the specialized-carrier concept (Docket No. 18920). As might have been expected from previous experience in the Computer Inquiry, the question elicited strong support for the MCI and Datran concepts from all sections, with the exception of the established carriers, of course.[85]

The Commission opted in favor of the specialized-carrier concept in order to obtain the new services and to stimulate better performance of the existing carriers. Its decision of May 1971 [Specialized Carrier] permitted virtually free entry of all financially and technically qualified applicants into the specialized carrier service, and it denied a joint petition by the National Association of Regulatory Utility Commissioners (NARUC) and the Utilities and Transportation Commission of the State of Washington (WUTC), for reconsideration of the specialized common carrier concept.[86] With the assistance of the Commission, competition was being established and the FCC could be expected to reject tariffs submitted by the established carriers which would tend to drive out their new competitors.[87]

---

[85] Ibid., p. 1267.

[86] U.S. Superintendent of Documents, 38th Annual Report/Fiscal Year 1972, Federal Communications Commission (Washington, D.C.: Government Printing Office), stock no. 0480-00271, p. 99.

[87] Stuart L. Mathison and Philip M. Walker, "Specialized Common Carriers," Telephone Engineering and Management (October 15, 1971), p. 58.

As late as May, 1974, however, the Commission was required to intercede in behalf of the specialized carriers [MCI in this case] and direct AT&T to provide interconnect services, which AT&T was attempting to deny (through subtle legal machinations) in its continuing fight to minimize the effects of the new carriers.[88] The Commission also placed all telephone companies on notice that its policy declarations applied to them as well as to Bell and emphasized that it expected compliance.[89]

In January, 1975, a three-judge tribunal of the U.S. Court of Appeals, in a unanimous decision, affirmed that the FCC had properly exercised its "broad" and "sweeping" authority in common carrier regulation to produce its specialized common carrier policy. This decision concluded the appeal that had been filed by NARUC/WUTC challenging the FCC's denial of its earlier petition. Judge Browning noted that the Commission's authority had been specifically stated broadly "to meet the needs of a dynamic rapidly changing industry."[90] While the telephone companies

---

[88]"AT&T Loses Motion: Will Reconnect MCI's Private-Line Services," Wall Street Journal (May 3, 1974), 17:4. Also see WSJ 1/4, 19:1; 2/19, 17:1; and 4/24, 15:1.

[89]"FCC Gives Green Light to Specialized Common Carriers," Microwaves (June, 1974), p. 24. An MCI official was quoted as saying, "It was a breached delivery, but at least we're born."

[90]"FCC Properly Exercised 'Broad' and 'Sweeping'

opposed this competition, they also recognized the inevitability in its development as was indicated by the nature of their responses and their various tariff revisions.[91]

Although the factors identified thus far were influencing both the regulated and the regulator during DOMSAT, there were additional influences on the regulator. They were to be found within the government structure and were evident during two separate administrations.

B. The Policy Makers

The influences that the Eisenhower and Kennedy administrations had on DOMSAT were manifested in the Communications Satellite Act of 1962, which laid the foundations for DOMSAT. But both the Johnson and Nixon administrations would play major roles before the DOMSAT policy was finalized, and an assortment of personalities would participate.[92] In technical

---

Powers," Telecommunications Reports, Vol. 41, No. 4, (January 27, 1975), pp. 6-9. Keller and Heckman, Telecommunications counsel for the National Retail Merchants Association, appeared before the San Francisco Court of Appeals to argue in favor of upholding the Commission's decision. See W. H. Borghesani, Jr., "The Evolving Telecommunications Regulatory Environment--Parts I and II," Telephone Interconnect Journal (November, 1973-January, 1974), pp. 20-22, 31 and pp. 22-24.

[91] Letter from Henck, May 17, 1976.

[92] Appendix B provides a picture in time of the relationships that existed between prime participants in DOMSAT.

circles the extent of political participation in policy making is well recognized, as was so aptly put by the editors of a recent technical society publication:

> The technical solutions to problems presented by user needs are constrained or inspired just as much (if not more) by economics and the impact of conflicting societal pressures (as embodied in regulatory decision) as by the cleverness or lack of cleverness of the engineer.[93]

From 1965 to 1968, the White House, under the Johnson Administration, turned its attention toward a review of the Nation's telecommunications policy. The President's Task Force on Communications Policy, appointed in 1967 and supported by the FCC, the National Academy of Engineering and others conducted extensive reviews of both policy and technology in a variety of areas with considerable emphasis on DOMSAT.[94] It recommended that a COMSAT-directed pilot program be authorized for domestic satellite, since at that time the potential benefits were too indeterminate to permit an informed decision as to how such satellites might best be utilized "in the public interest."[95]

---

[93]Paul E. Green, Jr. and Robert W. Lucks (eds.), Computer Communications (New York: IEEE Press, 1975), p. 4.

[94]Michael J. Morrissey and John J. Smith, "FCC Activities: Domestic Satellites", George Washington Law Review, 41 (May, 1973), pp. 746-759.

[95]President's Task Force on Communications Policy, Final Report (Washington, D.C.: U.S. Government Printing Office, 1969), Chapter 5, pp. 9-17.

However, in 1969, after a change in administrations, the questions of ownership and operation were reexamined and an "open entry" policy was recommended.

As noted in Chapter I, the Commission has the responsibility of supervising schedules of charges filed with it, for licensing new entrants, and for issuing certificates of public convenience. The agency can execute these functions by formulating policy either through rulemaking or through adjudication. Frequently the agency will issue a tentative decision or proposed rulemaking which is open to comment by interested parties to enable it to make an informed final decision of policy.[96] Adjudication on the other hand has been denoted as the agency process for formulation of a formal order. Congress has conferred on the U.S. Courts of Appeals jurisdiction to review only the final orders of the Commission. It is not uncommon to find reversed adjudicative proceedings, but the agency's judgment regarding rulemaking has seldom been disturbed.[97]

The purpose of the proposed rulemaking for Docket No. 16495 was "to facilitate expeditious actions on the applications [for DOMSAT systems] and prompt

---

[96] Taub, "Federal Communications Commission Regulation of Domestic Computer Communications," p. 951. Rulemaking is defined as the process of formulating, amending or repealing a rule.

[97] Ibid., p. 952.

attainment of the potential benefits of the satellite technology in the domestic field" and to keep open the proceedings of the docket.[98] No cutoff date for the filing of applications was established at that time. With proposed filings submitted, the Commission would then decide on the questions of policy presented. The Commission invited all those who were legally, technically and financially qualified, to file applications for satellite systems that proposed services directly to the public on a common carrier basis, or to other common carriers, or for a combination of such arrangements.[99] At the same time, the Commission also instituted a general rulemaking on the policies to be followed in the event of technical or economic conflicts between applications and on the initial role of AT&T in DOMSAT, access to Earth stations, and procurement policies.[100]

It would take two more rulings, a change of administrations, and more than thirty-three months before all matters were finally resolved.

1. Activities of the Johnson Administration

On August 14, 1967, President Lyndon B. Johnson,

---

[98] 22 FCC 2d 86, Report and Order, FCC 70-306. To be referred to as the First Report.

[99] Robert L. Werner, "A Lawyer Looks at our Communications Policy," Jurimetrics Journal, 11 (December, 1970), p. 89.

[100] 22 FCC 2d 810, Notice of Proposed Rulemaking.

provided the following thoughts to the Congress in a policy message:

> [T]he challenge is to communicate. No technological advance offers a greater opportunity for meeting this challenge than the alliance of space exploration and communication. Since the advent of the communications satellite, the linking of one nation to another is no longer dependent on telephone lines, microwaves or cables under the sea.[101]

He noted, "Communications satellites have domestic as well as international applications," and asked the Task Force, which he was appointing by this message, to look at the Nation's overall communications policy and to address the following two questions in its primary efforts:

> --How soon will a domestic satellite system be economically feasible?
>
> --Should a domestic satellite system be general purpose or specialized and should there be more than one system?

Additionally, the Task Force was to determine if either the Communication Act of 1934 or the Communication Satellite Act of 1962 required revision and to consider the international aspects of communication satellites and the utilization of the frequency spectrum.[102] It was clear in his message that the potential impact of

---

[101]U.S. Congress, Message from the President of the United States, House of Representatives, House Document No. 157, 90th Congress, 1st Sess., August, 1967, p. 1.

[102]Ibid., pp. 5, 8, and 9.

the communications satellite on domestic policy was well recognized and that this was one of the major reasons that overall communications policy was being examined.

While much of Johnson's message was in an international context, it would have been unreasonable to expect that the domestic communications environment would remain static while the international environment exploded with this new technology. As Johnson said:

> Such an archaic system of international communications is no longer necessary. The communications satellite knows no geographic boundary, is dependent on no cable, owes allegiance to no single language or political philosophy. Man now has it within his power to speak directly to his fellow man in all nations.[103]

a. The Johnson Task Force

The Task Force was chaired by Eugene V. Rostow, the then Under Secretary of State for Political Affairs. On December 7, 1968, after more than fifteen months of policy examination, the Task Force [with the cooperation of fifteen departments and agencies of the Federal Government] issued its Final Report.[104]

---

[103] U.S. Congress, Message from the President, p. 4.

[104] Final Report, President's Task Force on Communications Policy, December 7, 1968 (Washington, D.C.: Government Printing Office, 1969), 0-351-636. Those involved were State, Defense, Justice, Commerce, Labor, HEW, HUD, Transportation, USIA, NASA, Bureau of the Budget, Council of Economic Advisors, Office of Science and Technology, National Aeronautics and Space Council and the Office of Telecommunications Management. The

Noteworthy was the fact that this Task Force had attempted the broadest review of national telecommunications ever undertaken; also that one of the most significant problems identified involved the efficient introduction of communications satellites into domestic service.

The Report's staff studies indicated that satellites would be competitive with terrestrial facilities in meeting some domestic communications requirements.[105] The economic practicality was no longer considered to be a question. The Report envisioned complementing "overlaid" networks but not revolutionizing the fabric of the existing terrestrial systems. The satellite's ability to reallocate communications capacity flexibly and rapidly among a number of individual routes [in other words, "variable capacity routes"] was seen in the Report as possibly "economically and operationally attractive" but impossible to evaluate.[106] Permitting the competitive market to do this was not even suggested. The Saturday Review noted in retrospect [1971] that:

> Some observers think that, in 1967, the FCC would have designated AT&T

FCC's participation was ex-officio in nature, necessary because of the Commission's statutory responsibility.

[105] Ibid., Chapter 5, p. 7.

[106] Ibid., pp. 5-6.

> and COMSAT as the chosen entities [for DOMSAT]. Authorization was delayed pending the report of President Johnson's Task Force. . .[107]

But the article provided no support for this observation. However, even if the observation was accurate, the Task Force Report certainly didn't provide much in the way of options for the Commission to consider. The available data was believed to be insufficient.

Answering Johnson's questions of "how soon" and "what kind", the Task Force sidestepped the issues by noting that "a number of unresolved questions make it premature to establish full-scale domestic satellite operations at this time" and "substantial disagreement also exists as to the comparative advantages of a general purpose domestic satellite system and one or more systems 'dedicated' to specialized uses. . . ."[108]

The Task Force concluded by advocating no more than what the FCC was alleged to have been previously considering, that a COMSAT pilot program should be established first to gain insight. The Report noted:

> [W]hile our own independent estimates do not indicate that substantial economies will result in the very near term from the substitution of satellite facilities for a terrestrial equivalent, neither do they show that some uses of domestic satellites--particularly for

---

[107]Robert Lewis Shayon, "Bird Watching," Saturday Review, April 17, 1971, p. 57.

[108]Final Report, President's Task Force on Communications Policy, Chapter 5, p. 9.

> television distribution--are bound to be uneconomical.[109]

With this statement both sides of the coin were covered.

In fairness, though, the general themes of the Report [the needs for more competition and greater innovation] can be considered positive and constructive. It recognized that the Commission had broad regulatory powers under both the 1934 and 1962 Acts and recommended that they be strengthened, by Congress if necessary, to cope with the rapidly changing technological environment and to insure effective regulation.[110]

While its technical projections were limited, the Report did provide the following observation on message switching in a footnote:

> When teleprocessing [computercations] moves into a separate digital network, with time division switching, economies of scale may become very strong.[111]

In the Report's dissenting statement, Vice-Chairman James D. O'Connell added emphasis to a point that was only lightly touched upon in the basic report. Expressing

---

[109] Ibid., pp. 17-18. ABC had made that same appraisal in 1965.

[110] Werner, "A Lawyer Looks at our Communications Policy," p. 81. Congress took no action in this area.

[111] Final Report, President's Task Force, Chapter 6, p. 33. Someone at IBM must have certainly read and pondered that thought for a while, as well as COMSAT, AT&T and many others. The Task Force, however, appeared to have no appreciation at that time for the potential of satellite technology or for the economies that digital technology could bring.

confidence in the regulatory process,[112] O'Connell felt that the Executive Branch and the FCC could initiate the necessary improvements called for in the area of telecommunications policy. But he emphasized that a substantial augmentation of the resources of these offices would be necessary to accomplish these improvements for the task was enormous.[113] Although the Task Force Report was widely quoted, it was never released officially, nor was it acted upon by either the outgoing or the incoming administration.[114]

b. The National Academy of Engineering

Some of the technical considerations of DOMSAT were investigated by the National Academy of Engineering (NAE) and the results were furnished to the Task Force in a separate report.[115] The NAE had been established in 1964 to share in the responsibilities of the

---

[112] It must be kept in mind that the goals of regulation and competition are identical--efficiency, progressiveness, reasonable prices and satisfaction of diverse needs. See Lionel Kestenbaum, "The Limits of a Regulated Monopoly: Telephone Attachments, Interconnections, and Use of Circuits," The Antitrust Bulletin, 13 (1969), pp. 979, 983.

[113] Final Report, President's Task Force, Tab B, pp. 7-9.

[114] David C. Acheson, "Domestic Satellite Developments," Public Utilities Fortnightly, 86 (September 24, 1970), p. 68.

[115] Reports on Selected Topics in Telecommunications, The Final Report by the Committee on Telecommunications, National Academy of Engineering (Washington, D.C.: National Academy of Sciences, 1969).

National Academy of Sciences. The Committee on Telecommunications had been established in March 1968 to (1) undertake preliminary review and evaluation of some of the reports of the Task Force and (2) to act in an advisory capacity to the Task Force on technical matters.[116] The NAE's intent was to contribute further understanding of selected problems in order to encourage action towards their solution. It was not concerned with organizational or jurisdictional considerations. Rate-policy considerations as they affected the revenue to be expected from a given set of communications facilities were also excluded.

The NAE compared satellite and terrestrial systems, making observations on demand projections, economic considerations and system designs. Because of land cable improvements, the cost advantages of microwave relay for large cross-section routes had been disappearing since 1948. The leading candidates for the larger cross-section land routes of the future were millimeter-wave waveguide systems and laser guided-beam systems. Both were extensions of the land cable concept, being fully guided systems. In making comparisons between alternative systems it is important to realize that it was being done at a particular instant in time for particular demand and

---

[116]Ibid., pp. 1-4.

technology predictions. Longer-range views require that added attention be given to the uncertainty that exists with respect to future developments.

For a realistic analysis, two specific comparisons of satellites and cables were done, based on a simplified model of long-haul transmission in the Atlantic basin for the period 1968 to 1985, starting with the 1967 demand projections of the Joint World Plan Committee of the ITU.[117] It was found that all new demand could be met with satellites, that such a system would be a minimum cost system, that a premium must be paid for the use of cables, and that cables could be used to temporarily postpone an investment in satellites for a year or two in certain cases.[118]

As seen by the NAE, the economic considerations of a domestic satellite system tended to focus on the calculation of a break-even distance above which the satellite service would be cheaper than terrestrial links. Domestic service meant the use of satellites more or less within the boundaries of a particular nation. The break-even distance is influenced strongly by the extent of terrestrial facilities

---

[117] International Telecommunication Union, Joint World Plan Committee of the CCIR and CCITT, General Plan for the Development of Interregional Telecommunication Network, 1967-1970-1975 (Geneva: ITU, 1968; meeting held in Mexico City, 1967).

[118] Reports on Selected Topics in Telecommunications, pp. 5-8, 13.

already in existence at the time the new satellite service is contemplated. Break-even distance is a complex function of the total traffic handled and the number of routes to be served (route traffic density). The space segment cost of a satellite system is almost independent of whether the total traffic capacity of the satellite is used in connecting a high traffic-density route or in interconnecting many points with lower traffic density. In the latter type system the break-even distance could be lower than in the former but Earth station costs are a critical factor. The problem would be simplified if the only factors involved were careful forecasts of the service requirements and the economic optimization of the technical approach. However, matters of national prestige, policy determination regarding the scope and purpose of service, and political influence were also in the picture and had to be taken into consideration and included in the NAE's report.[119]

The NAE recognized that satellite systems had a distinct cost advantage over undersea cable and provided flexible low-cost service through multiple-access capabilities. Because of this, if traffic demands were small, the alternative of sharing the international satellite with domestic traffic clearly

---

[119]Ibid., pp. 21-22.

offered the most economic solution. On the other hand, if traffic demand was large (thousands of channels) a separate satellite was thought to be more practical. However, satellite design (to include replacement and growth-in-capacity considerations) could be more effectively optimized if only domestic service was to be provided. One result of rather conservative technology forecasts throughout the study was that costs-per-channel of the systems compared were thought to be reasonably correct for 1980 even though the actual technology in that time period may be different from what was projected.[120] The satellite technology available (as seen by the Academy) offered attractive prospects for domestic communication services on either a modest scale or on a large scale and new types of service which would otherwise be impractical would "no doubt lead to very serious consideration of this type of facility."[121]

2. A Change of Administrations

As Presidents changed, so did perspectives on policy. The FCC took no action on the domestic satellite

---

[120]Ibid., pp. 19-20, 23. In the DOMSAT filings, there would be applications for 24 orbit slots, but four of these were designated for ground spares. Even with Canada's proposed satellites included, all systems could be accommodated on an individual basis.

[121]Ibid., pp. 25-26.

issue while waiting for an expression of opinion from the new Administration.[122] Indications that the Task Force Report was unacceptable to the new Nixon Administration and that the White House would support an intensive reexamination of national telecommunications policy (and the implied postures toward competition and regulation in the communications industry) had been in the air since Nixon's election in 1968.[123]

In testimony before the House Subcommittee on Communications on March 6, 1969, Congress had been made aware of the lack of activity on the DOMSAT issue and the Final Report of the Rostow group. The FCC's Chairman Rosel Hyde noted:

> We deferred reaching any policy decisions in our domestic satellite inquiry pending conclusion of the work of the President's Task Force on Communications Policy [Rostow's Task Force]. . .Our staff likewise participated in discussions and in my judgment was very helpful in providing information to the task force. So, we are conversant with their deliberations, notwithstanding the fact that the report which was submitted to the Chief Executive has not been released.[124]

---

[122] Acheson, "Domestic Satellite Developments," p. 68.

[123] Thomas P. Murphy, "Federal Regulatory Policy and Communications Satellites: Investing the Social Dividend," <u>The American Journal of Economics and Sociology</u>, <u>31</u> (October, 1972), p. 340.

[124] U.S. Congress, House, Subcommittee on Communications and Power, <u>First Session on the Jurisdiction and Activities of the Federal Communications Commission</u>, 91 Congress, March 6, 1969, pp. 3-4.

When the subcommittee's Chairman, Torbert Macdonald, inquired as to the whereabouts of Rostow's study, Hyde replied:

> It is my understanding that the Chairman of the task force, which was Mr. Rostow, transmitted the report to the President. This would be President L. B. Johnson and I presume it would have been transmitted from his administration to the new administration. . .I believe it was in November but I can get the date for you.[125]

The following information was submitted for the record in the hearing report:

> The date of the letter of transmittal of the President's Task Force Report on Telecommunications was December 7, 1968.

There were new forces at work in the White House and the questions of satellite communications policy were being escalated to the highest macropolitical level. A small working group was formulating "Administration suggestions" regarding the introduction of communications satellites into U.S. domestic communications. A White House announcement, dated July 22, 1969 stressed: "We will be concerned, of course, with the general structure and direction of the industry and not with specific applications pending before the Commission." Among the membership of this working group was special Presidential staff assistant, Dr.

---

[125] Ibid.

Clay T. Whitehead.[126] The lucrative aspect of the DOMSAT market seemed to draw interest also. As COMSAT President Joseph V. Charyk observed:

> I think the very success of satellites for international application has complicated the problem of authorization on the domestic scene, because with the international success, it would appear that satellites are a good thing, and so everyone wants to get into the act domestically.[127]

With the release of a January 23, 1970 Memorandum from Presidential Assistant Peter Flanigan to the new FCC Chairman Dean Burch, the Nixon interest in DOMSAT was confirmed:

> Federal policy on domestic satellite communications has been long delayed. . . At this stage of domestic satellite planning, it is not possible to identify major economies of scale. Rather, it appears that a diversity of multiple-satellite systems as well as multiple-earth stations will be required to provide a full range of domestic services.
>
> Further we find no public interest grounds for establishing a monopoly in domestic satellite communications.

It disagreed with the Task Force Report, recommending an "open skies" policy:

> Government policy should encourage and facilitate the development of

---

[126] Kurt Borchardt, Structure and Performance of the U.S. Communications Industry (Boston: Harvard University Press, 1970), pp. 129-130.

[127] Murphy, "Federal Regulatory Policy," p. 344; see also House Report No. 859, Assessment of Space Communications Technology, note 131 infra.

> commercial domestic satellite communication systems to the extent private enterprise finds them economically and operationally feasible.[128]

a. The Commission

Although the FCC announced its Proposed Rulemaking and First Report on DOMSAT less than sixty days after this memo was released, it should not be quickly interpreted as a direct reaction to the Flanigan memorandum, for the people on the Commission at the time were not the type to jump at an Executive dictate, particularly Nicholas Johnson, a veteran Commissioner since 1966 who was known for speaking his mind. He had described the period following the Carterfone decision in the following manner:

> We haven't the slightest notion where we're going (or, indeed, even where we want to go)--but we know we're getting there a whole lot faster than before.[129]

He paints a slightly different picture of the DOMSAT atmosphere at the time. Just one week prior to the

---

[128] 22 FCC 2d 0125, The White House, Memorandum for the Honorable Dean Burch, Chairman of the Federal Communication Commission, January 23, 1970.

[129] Nicholas Johnson, "Harnessing Revolution: The Role of Regulation and Competition for the Communications Industries of Tomorrow," The Antitrust Bulletin, 13 (Fall, 1968), p. 881. Johnson has been described by Dr. Thomas Murphy as "a man who must seem inscrutable to some while appearing most open and logical to others. As one who champions the public interest, he has made powerful enemies in the industry." See note 123, supra, p. 342.

First Report, Johnson was quoted:

> [I]n the summer of 1969 the Nixon Administration asked the FCC to delay once again while another Executive Department review was undertaken.[130]

A House Report on space communications also indicates that the FCC was on the verge of announcement in its decision regarding the domestic [satellite] system about July 1969, when an additional delay was brought about by an announcement that the White House had established still another task force, chaired by Clay T. Whitehead, to further review policy on domestic satellite service.[131]

The Flanigan memorandum was interpreted by some observers as signalling a conscious effort on the part of the Executive to take a more forceful initiative in planning and advocating measures that fall within the purview of the regulatory agencies.[132] Speaking

---

[130]Remarks by Nicholas Johnson, March 12, 1970, "The Capacity to Govern: The Role of the FCC in the Development of National Policy for Computer Communications," Hopkins-Brooking Lecture Series, Computer Communications and the Public Interest, Advanced International Studies.

[131]U.S. Congress, House, Committee on Science and Astronautics, Assessment of Space Communications Technology, H. Rept. No. 859, March 3, 1970, 91st Congress, 2nd Session, p. 19. The report concluded that DOMSAT had not been inhibited thus far by industry institutions [COMSAT had been advocating a domestic satellite for the U.S. for almost 4 years (p. 4)]. There was no assessment as to whether government institutions had hindered DOMSAT.

[132]Acheson, "Domestic Satellite Developments," p. 68.

before the American Bar Association in August 1970, COMSAT's General Council, David C. Acheson, saw the memo this way:

> [O]ne may wonder whether the January 23rd memorandum represents a truly sophisticated view of the domestic satellite question. "Competition" is an attractive word, but not every policy carrying the name is necessarily the real goods, union-made, preshrunk and Sanforized. For the near future there are probably only two uses of domestic satellite that would support the large requisite investment: They are the domestic telephone network and commercial broadcast distribution. . .[133]

At the FCC both old forces and new forces were at work. Besides the "public interest" minded Johnson, there was the Common Carrier Bureau's Bernard Strassburg, its chief since 1964, a veteran of more than twenty-five years with the Commission, and one who had for years recognized the importance of data transmission.

The Commission also had a new boss, Dean Burch. Burch, the first Republican to chair the Commission in nearly a decade, was also a strong advocate of the public interest. He and Johnson, his most flamboyant democratic colleage, contributed much to the formulation of DOMSAT policy. With the appointment of Burch as

---

[133]Ibid. Even COMSAT had not recognized the potential of the computercations market, even though it had small processors conversing via satellite at the time. See the discussion of SPADE in Network Developments, Chapter IV.

Chairman the pace of activity and interest picked up considerably.[134] Burch, who ran the 1964 presidential campaign of Barry Goldwater, was known as a champion of free enterprise, a "laissez faire" capitalist who would let the marketplace regulate communications. His detractors called him a "reckless intellectual hipshooter" and believed him to be "a bit rash".[135] One of the political people, he is listed as a Conservative from Arizona, but he is considered much broader and smarter than the connotations of any label. Within the Commission, he became known as the most independent of the group, believing that big communications was as dangerous as big government and big labor.[136] The Commission's agenda was already loaded with blockbuster issues:

--the ownership of and the number of domestic communications satellites and what services they would offer;

--the future of digital communications; and

--the entry of the special service common carriers into the market who planned to compete with AT&T.

Burch had his work cut out for him.

---

[134] Murphy, "Federal Regulatory Policy," p. 341.

[135] Lois Vermillion, "Dean Burch: FCC's Pragmatic Boss," Electronics (September 28, 1970), p. 85.

[136] Jimmy Breslin, How the Good Guys Finally Won New York: Ballentine Books, 1976), pp. 132-133. See note 157, infra.

b. The Office of Telecommunications Policy

The Office of Telecommunications Policy (OTP) was a new addition to the FCC's environment that needs to be noted. In February 1970, President Nixon submitted a reorganization plan to Congress to establish within the Executive Office of the President an Office of Telecommunications Policy with the specific objective of representing the President's views to the FCC and Congress regarding matters such as CATV [cable television], regulation of pay television, assignment of scarce portions of the radio spectrum, "diversification of media ownership. . .and the encouragement of competitive challenges against the Bell Telephone System in specialized communications services."[137] With no objection from Congress, it would become effective within 60 days "equal in rank with the President's staff panels on the economy, science and environment" and it would be headed by Dr. Whitehead.[138] The Executive's Office of Telecommunications Management and Nixon's Task Force were transformed into the OTP in March and given broad powers to shape government policy on computers and communications.[139]

---

[137]Thomas P. Murphy, "Technology and Political Change: The Public Interest Impact of COMSAT," The Review of Politics, 33 (July, 1971), p. 423.

[138]Ibid.

[139]Murphy, "Federal Regulatory Policy," p. 342.

What was the philosophy underlying this new approach? Was there to be a change in policy resulting from the OTP's formation? It is easy to understand that there might be a difference of opinion as to whether the monopoly regulated should be a public monopoly, a private monopoly, or a hybrid organization such as COMSAT. The private carriers had been shunted off the stage in 1961 and 1962 by the Democratic Administration after securing statements from the Eisenhower Administration favorable to exclusive control of communications satellites by private industry. However, the FCC had favored neither COMSAT nor the carriers in their rulings and there was legislative history that supported these decisions.

A general review of the White House statements will clearly show that the Nixon philosophy was not only intended as supporting greater competition in communications but in other fields (for example, railroads and aviation) as well.[140] This new office was not intended to usurp the prerogatives or functions assigned to the FCC by Congress; it was believed that this new department and the Commission would cooperate in achieving reforms in the telecommunications field.[141]

---

[140] Murphy, "Technology and Political Change," p. 424.

[141] Taub, "FCC Regulation of Domestic Computer Communications," p. 953.

c. Congressional Oversight

In addition to its legislative power, Congress has at its disposal a variety of subtle techniques which can influence Commission decisions. One of them is legislative oversight. Standing committees exercise "continuous watchfulness" of the execution of laws by the administrative agencies under their jurisdiction.[142] The standing committees that subpoena the FCC have a significant impact on its activities. "No other Federal agency has been the object of as much vilification and prolonged investigation by Congress," notes Erwin Krasnow, a member of the FCC bar. During the Ninety-First Congress over twenty-five committees and subcommittees attempted to oversee virtually every aspect of the FCC's activities.[143] Congress was interested in DOMSAT.

In 1971 oversight hearings, Congressman Macdonald expressed deep concern over the slow progress that the Commission had been making in the area of domestic satellites.[144] Chairman Burch, in his prepared

---

[142] George B. Galloway, "The Operation of the Legislative Reorganization Act of 1946," American Political Science Review, 45 (1951), pp. 59-60.

[143] Taub, "FCC Regulation of Domestic Computer Communications," pp. 952, 962.

[144] U.S. Congress, House, Subcommittee on Communications and Power, First Session on the Jurisdiction and Activities of the Federal Communications Commission, 92nd Congress, April 29, 1971, p. 1.

statement pointed out:

> Satellites represent an entirely new potential for domestic communication. Depending on the direction and scale of its development, domestic satellite communications may affect many if not all terrestrial systems and existing uses of radio and cable.
>
> No system can be constructed without Commission authorization. In this sense, the Commission is writing on a clean slate. At the same time, however, we are aware of the interrelationship between any satellite system and existing terrestrial facilities. Our decisions in this area will necessarily have substantial economic and technological impact all across the common carrier area.
>
> The Commission decided in March 1970 that satellites offer sufficient promise as a mode of domestic communications to warrant assignment of frequencies and use of orbital parking spaces.
>
> The first application for a domestic satellite system was filed on July 30, 1970 and we now have pending before us 10 applications--eight for complete systems and two for earth stations only.[145]

DOMSAT had been seen by MCI, the specialized carrier, as advantageous for supplementing their developing terrestrial system. Four areas of potential impact [(1) improved network management, (2) heightened total system reliability, (3) improved earth station flexibility, and (4) increased broadband capacity]

---

[145]Ibid., pp. 9-10.

were seen.[146] Consequently, MCI Lockheed Satellite Corporation, a joint venture by MCI with the aerospace giant, had announced a week prior to March 15, 1971 [the deadline that the Commission had finally set for DOMSAT construction applications] that it was applying for a permit to construct a $168 million advanced-technology DOMSAT for operation by 1975.[147] This is the root of the story that evolves in Chapter IV.

When Senator Pastore opened oversight hearings of the FCC in February, 1972, his general comments throughout the hearings seemed sympathetic to the complexity of the Commission's task in the face of the different points of view that the public interest can take.[148] Chairman Burch explained that the domestic satellite issue had been delayed by the priority that had been given to the cable television issue. Although open entry appeared favorable, the real problem was trying to determine who should not be permitted into the business.[149]

---

[146]Tom Leming, "An Appraisal of the Role of Satellites in Domestic Communications," presentation before the AIAA 4th Communications Satellite Systems Conference (Washington, D.C.: April 14-16, 1972).

[147]"Lockheed, MCI Team to Propose Domestic Satcom Plan for 1975," Aviation Week and Space Technology (March 18, 1971), p. 186.

[148]U.S. Congress, Senate, Subcommittee on Communications, Overview of the Federal Communications Commission, 92nd Congress, February 1 and 8, 1972, pp. 2, 88-90.

[149]Ibid., p. 98.

Commissioner Johnson, in his last official appearance before Pastore's subcommittee the following year, attempted to identify why the Commission generally had difficulty in developing communications policy:

> There is much concern these days, as well there should be, about the power of Congress. The relationship of the independent regulatory agencies to the Executive Branch and the Congress is a useful arena in which to examine that power. The idea of the FCC as an independent agency established as "an arm of Congress" and responsible to it is a creed to which the FCC genuflects on all appropriate occasions before Congressional committees. But it is a creed that no one seriously believes any longer, because it is not true. . . the Executive Branch can and does exercise control over the FCC.[150]

Additional dialogue went as follows:

> Senator Pastore. The question, Mr. Johnson, is why are you not getting them [decisions] out? Whose fault is it?
> Mr. Johnson. I think it is a variety of factors. I think part of it is that while we have made some progress in terms of a policy planning office--we now at least have an office, even if we only have one man in it--we, basically, have never really had the resources to do policy analysis that come anywhere close to what Dr. Whitehead has, or the Defense Department has, or even the FAA, which used to have a budget for communications research that exceeded the FCC's entire budget.[151] We just never

---

[150]U.S. Congress, Senate, Subcommittee on Communications, Overview of the Federal Communications Commission, 93rd Congress, February 22, 1973, p. 62.

[151]Burch believes that the rethinking of the

> came close to having the number of professional people we need as economists and whatnot to help us with some of these policy questions.[152]

During the same hearings, Commissioner Richard Wiley added:

> I think the fact that there are more issues, yet to be resolved is simply a reflection of the fact that the technology of the industries that we regulate is running far ahead of the policy and the law.[153]

To what extent Congress influenced DOMSAT is questionable, although certainly in this manner, some practical check on regulatory action was undoubtedly being imposed by Congress. Congressional approval or disapproval of agency action is frequently intimated and sometimes expressed, either by individual Congressmen or in committee reports. Such expressions have some influence on regulatory action, although the degree of influence varies greatly depending on a variety of factors. Congress is, of course, in a position to make its views effectively felt by regulatory agencies, either by passing controlling statutes or by pulling on the purse strings of

---

problem, done by Tom Whitehead and a group of White House experts, significantly influenced the adoption of an "open skies" policy. Letter from Dean Burch, Pierson, Ball, and Dowd, Washington, D.C. to Robert S. Magnant, May 12, 1976.

[152] Overview of the FCC, 1973, pp. 65-66.

[153] Ibid., p. 69.

appropriations. Both of the approaches are highly effective but infrequently invoked.[154]

Other Congressional testimony will be noted throughout the final DOMSAT proceedings in the next section. However, the following points found in a recent House Report are considered worthy of addressing here, for they put the Commission's general problem of policy formulation into perspective. In July 1975, the House Subcommittee on Communications held three days of hearings on Telecommunications Research and Policy Development. Two of the findings of the interim subcommittee report are noteworthy to this review:

> (1) Given the budgetary constraints under which FCC operates, it appears that the Commission will find it extremely difficult to conduct meaningful telecommunications research and policy development internally, although new efforts might result from the recent attempt to enlist the support of the research community.[155]
>
> (2) What makes the problem of meaningful regulation and policy development even more difficult is that electronic and telecommunications technologies appear to be on a competitive collision course. There is a distinct promise--some would say threat--of the giant in computer technology, IBM, competing with the giant in telephone technology, AT&T,

---

[154]Lee Loevinger, "Regulation and Competition as Alternatives," The Antitrust Bulletin, 11 (1966) p. 128.

[155]U.S. Congress, House, Subcommittee on Communications, Interim Report and Recommended Courses of Action Resulting from the Hearing on Telecommunications Research and Policy Development, 94th Congress, December 1975, p. 11.

> and the giant in postal communications, the U.S. Post Office. All have a stake in the future of telecommunications along with COMSAT and the communications satellite manufacturers, the broadcasters and the cable operators, among others.[156]

3. The Final Proceedings

Seven years of deliberations, hearings and discussions were slowly drawing to a close. The DOMSAT policy that was finally reached was primarily a product of political reaction to a new invention in a field that was dominated by one private company on the ground and another one in the sky. The policy-making arms of the government perceived, rightly or wrongly, a public distrust for bigness, including government. Under the safe banner of competition as a means to stimulate new ideas and reduced costs, the Commission authorized basically an "open skies" policy as a means to contain bigness.[157]

Burch saw the DOMSAT issue as an example of unnecessary delay. "[W]hen a technology is ready, it shouldn't be kept in the wings," he believed and was inclined to think of the FCC's role in negative terms as "not imposing artificial barriers to technological development." Ironically his favoring of a competitive

---

[156]Ibid., p. 15.

[157]Letter from Mr. Richard P. Gifford, Vice President, Communications Projects, General Electric to Robert S. Magnant, May 7, 1976. Emphasis added.

communications environment put him on the side of the Democrats on the Commission.[158]

By March 1971, with the filings of thirteen applicants before it (five for partial systems), the FCC had some decisions to confront (i.e.--Open entry? Who owns Earth stations? Who can have access?). But after reading the staff's summaries and recommendations, the commissioners found that they could not reach a decision. It is easy enough to see why. A new technology was involved; many conflicting corporate and political interests were involved. The size of the prospective market was uncertain; indeed, since different companies had different notions about the best ways to exploit communication satellites, it might be more appropriate to talk about propspective markets, in the plural.[159]

The staff feared that open entry would not work and that it might end up in monopoly. The main difficulty was the enormous amount of capacity that would come into being if all proposed systems were built; the effect might be that most systems would operate at a heavy loss and only AT&T, which had no problem with a fill factor (because of its ready

---

[158] Vermillion, "Dean Burch: FCC's Pragmatic Boss," p. 85.

[159] John McDonald, "Getting Our Communication Satellite Off the Ground," Fortune, 86 (July, 1972), p. 66; see also the comments by Henson, note 202, infra.

market) would be able to survive.[160]

In October a "Dean Dean" letter from Tom Whitehead stirred the pot:

> There are customers waiting for satellite services and prospective suppliers with capital and the will to offer them on a commercial basis. We see no reason for the government to continue keeping these groups apart. No further studying of applications or enforced commercial arrangements would be as constructive for the using public or for the industry as the prompt opening up of this promising field.[161]

Whitehead made clear not only the Administration's impatience with the Commission to conclude the six-year-old proceeding but also its view that its own open-entry plan should be adopted. Such thinly-veiled criticism of the Commission's pace on DOMSAT seems slightly out of place when White House intervention had stalled progress twice before (the Johnson and Nixon Task Force Studies).[162]

When Senator Pastore opened oversight hearings in February 1972, he stated:

> Ultimately responsibility for the communications of the country rests with Congress; and if the agency we have created to carry out

---

[160]Ibid., p. 122.

[161]"Whitehead Prods FCC on Satellites," <u>Broadcasting</u> (November 1, 1971), p. 38; see also "FCC Expected to Permit Open Satcom Competition," <u>Electronic News</u> (December 13, 1971), p. 28.

[162]Ibid.

> the day-to-day implementation of the policies we have legislated is having difficulty, we want to know about it, and help if we are able. . . . When may we expect definitive Commission action on a domestic satellite policy?[163]

During the subsequent dialogue, Burch raised what appears to many to be the basic regulatory question that was addressed by the Commission; whether the domestic satellite technology would be given exclusively to the established carriers or whether it would be opened up for development by a number of competitors:[164]

> Mr. Burch. . . .The big problem. . . is the question of disqualification, not qualification. In other words, is there somebody who should not be in the domestic satellite business?
> Senator Pastore. I do not want to be misunderstood. I am merely trying to get a date Can we expect an answer before election day?
> Mr. Burch. Yes, sir.
> Senator, let me say this. Even if we acted tomorrow, the companies involved will then go back and restructure their applications and I cannot promise you there will be a domestic satellite beaming election results because that--I just do not know.
> Senator Pastore. But we will have a decision by then?
> Mr. Burch. Yes, sir. You will have a decision.

---

[163]<u>Overview of the FCC</u>, February 1 and 8, 1972, p. 2.

[164]Letter from Mr. Andrew Margeson, Staff Economist, House of Representatives.

> Senator Baker. Do you have any idea, Mr. Chairman, or can you tell us whether that decision will come in terms of the open entry concept or not?
> Mr. Burch. That will be determined as part of our determination.
> Senator Baker. Before the election?
> Mr. Burch. Within 60 days. Yes. Open entry sounds wonderful, but it does not solve the question assuming that open entry that means everybody or that there are certain people you keep out. That is the problem.[165]

a. Conflicting Opinions

The Proposed Second Report and Order, prepared by the Chief of the Common Carrier Bureau and based on comments received from the eight applicants and others, was released in March of 1972. This Report recommended that all applicants found qualified and proposing similar satellite technology should be required to consolidate their efforts in a partner-type relationship.[166] The staff felt that this grouping of applicants would reduce the required investment and overhead while leaving each entirely free to innovate.[167] This fell short of the open-entry scheme the White House had been advocating since January 1970. Commission officials indicated that, in view of the extensive analysis that the staff had

---

[165] Overview of the FCC, p. 98.

[166] 34 FCC 2d 38.

[167] Taub, "FCC Regulation of Domestic Computer Communications," p. 981.

written to support its position, the staff's position might be difficult to alter even though some Commissioners, including Chairman Burch and Richard E. Wiley, were said to favor a policy closer, though not identical, to the plan suggested by the White House. At a press conference, Bureau Chief Strassburg noted, "We have certain recommendations from Mr. Whitehead as we have from others. I'm not here to defend our proposals against the White House recommendations."[168]

The staff recommendation was designed to balance the benefits to be obtained from the competition of an open-entry plan--including the development of new technologies--against the financial risks of permitting unrestricted entry at a time when the market for domestic-satellite service was, in the bureau's view, limited, though growing. Whitehead's reported reaction was an intent to seek legislation if the staff's plan was adopted. He described "limited open entry" as a euphemism for saying that the Commission will decide what applicants are approved. "We would like completely open entry," he emphasized.[169]

OTP had no authority to direct the Commission to follow its lead. Legally, it had no more standing

---

[168] "FCC Staff Splits with OTP over Satellites," Broadcasting (March 20, 1972), p. 37.

[169] "Whitehead Flexes OTP's Muscle," Broadcasting (April 17, 1972).

in a Commission proceeding than any other petitioner. But it did have the influence of the White House behind it, and if that was not enough, it could, if it thought an issue important enough, go to Congress or to the courts to obtain Commission compliance.

And in saying that OTP would go to Congress in connection with the satellite matter, Whitehead said he considered the question an important one, and not only in its own terms. The form DOMSAT would take would provide a precedent for other forms of communications. He noted, "We ought to structure it right."[170] Oral arguments before the Commission were set to begin on May 1. After two days of testimony, it was still not clear whether the Commission was moving towards the position advocated by the Administration or not. An assortment of arguments had been presented. Justice, for instance, suggested that, during the initial phase of domestic satellite system operations, both COMSAT and AT&T be prohibited from transmitting television signals. The staff would have gone further with respect to AT&T and limited it to the carriage

---

[170]Ibid. Richard P. Gifford, General Electric Vice President, has taken the position that switched telephone service must be regarded as a "natural monopoly" and that the long-haul transmission "pipes" must also be considered "natural monopolies". He contends, however, that the services provided over these "pipes" enjoys no such status and recommends that the ownership be opened to major users as well as service suppliers. See "Noted on the News," Telecommunications Reports, 40, No. 39 (September 30, 1974), p. 37.

of its monopoly services--message toll and wide area telephone service.

The range of options discussed by the Commission appears to have been quite broad. At one end, Commissioner Nicholas Johnson was said to have expressed as his first choice the removal of AT&T and COMSAT from consideration for satellite business; other commissioners, however, reportedly appeared to favor the Justice-OTP approach.[171]

The concern of companies who sought entry into the new domestic communications satellite field but feared the consequences of unrestricted entry was expressed by a number of those who appeared before the Commission. Michael Bader, counsel for MCI-Lockheed, said:

> We couldn't afford open entry.
>
> Let qualified applicants in. But put some restrictions on so that there will be entry by companies other than AT&T and COMSAT. Don't give us the reward of a license that will be worthless if AT&T takes all the business. Private line business is petty cash to them. But it's lifeblood to us.[172]

On the other hand, COMSAT's David Acheson warned that the restrictions the staff has proposed putting on COMSAT--principally the requirement that it be forced

---

[171]"Is FCC Moving to Advocacy of Open-Entry Satellite Plan?" Broadcasting (May 8, 1972), p. 53.

[172]Ibid.

to choose between leasing facilities to AT&T and serving other customers--would seriously affect its role in the international field:

> If you tell COMSAT it is limited to any market. . .you'll lose a resource created by Congress. We ask the Commission to reject the restrictive policies recommended by the staff and to adopt an open-entry policy.[173]

b. Competition Reemphasized

Commissioner Abbott Washburn sees any attempt at identifying the most significant factor influencing DOMSAT policy as an invitation towards oversimplification:

> . . .the real policy problems are always multi-faceted and the pressures and voices influencing government are always many and varied. Problem solving in a democracy is never simple. But despite its frustrations and delays it is still the best method yet devised by man.[174]

However, the free-enterprise preferences of most of the Commissioners did raise the flag of competition once again and a policy of "multiple entry" for domestic satellites was adopted; but not without considerable dissention. When the Commission issues its Second Report and Order on June 16, 1972, it emphasized that

---

[173]Ibid.

[174]Letter from Commissioner Abbott Washburn, Federal Communications Commission, to Robert S. Magnant, June 3, 1976.

the true extent of public benefit was still undetermined.[175] In its attempts to insure effective entry into DOMSAT, the Commission could not ignore the economic dominance of AT&T, with its multi-billion dollar investments and operations, nor could it fail to consider the technical advantages enjoyed by COMSAT by virtue of its established experience and expertise.[176] In view of all the conflicts and complexities involved in the creation of this new industry, it is not surprising that the Commission's vote on the Second Report was four to three, or that the majority expressed great uncertainty as to how satellite communication would actually work out in practice in several different markets.[177]

In ruling for "multiple entry", the Commission said that it generally agreed with its staff's concern about AT&T's "strength and dominance", and about cross-subsidies that might eliminate competition at the outset; the majority also agreed to limit the relationship between AT&T and COMSAT. But the Commission disagreed with the staff's effort to solve the problem of fill by proposing coalitions based on

---

[175] 35 FCC 2d 844. Second Report and Order. To be referred to as the Second Report.

[176] Michael J. Morrissey and John J. Smith, "FCC Activities: Domestic Satellites," George Washington Law Review, 41 (May, 1973), p. 752.

[177] McDonald, "Getting our Communication Satellite Off the Ground," p. 126.

similar technologies. Instead, the Commission authorized all the qualified parties, acting either independently or in coalitions, to operate communication satellites. Burch joined Commissioners Wiley and Charlotte Reid in dissenting to the Second Report. In his words:

> We fully supported the concept of multiple entry (we had a hand in its invention) but we could not go along with some of the onerous conditions placed on AT&T and COMSAT on the basis of their past successes in other operations.[178]

However, Burch felt that the Commission "would have been well advised to adopt a posture of 'least is best,' to build a base from the irreducible marketplace realities (AT&T traffic) and to offer all applicants a maximum of options." Instead he charged that the Commission had "violated every one of these counsels of caution."[179] His concern was with the Report's two "fatal flaws: it may retard the evolution of satellite technology, not get it going, and it may thus withhold realistic benefits to the public."[180] He felt that the Commission could do better:

---

[178] Dean Burch, "Public Utility Regulation: In Pursuit of the Public Interest," Public Utilities Fortnightly, 92 (September 13, 1973), p. 72.

[179] Victor Block, "Burch Slams Commission on Domestic Satellite Decision," Telephony, 183 (July 24, 1972), p. 12.

[180] Ibid., p. 13.

> I believe that some inhibitors were called for as an interim measure--to give AT&T's potential satellite competitors a chance to get started. . .[181]

In the Second Report, COMSAT was required to form a separate corporate subsidiary if it wished to engage in any DOMSAT venture. The Report also prohibited any joint AT&T-COMSAT proposals. These restrictions were based primarily on the fact that AT&T owned 29 percent of COMSAT's stock and had the ability to elect three of COMSAT's 15 directors.[182]

The effective date previously set for the order (July 25) was postponed by the Commission until October, pending statements of intention from each of the applicants (with regard to their filings and intended courses of action). In the meantime, the Second Report became the subject of several petitions for reconsideration filed by AT&T, COMSAT and others and the Commission planned a later meeting to consider the docket again.[183]

On December 22, 1972, the Commission released its final report, a Memorandum Opinion and Order[184]

---

[181] Burch, "Public Utility Regulation," p. 72.

[182] Morrissey and Smith, "The FCC Activities," p. 753.

[183] David C. Acheson, "Domestic Satellite Proceeding--Status Report," Public Utilities Fortnightly, 90 (December 7, 1972), p. 53.

[184] 38 FCC 2d 665, Memorandum Opinion and Order. To be referred to as the Final Report.

and ended seven years of deliberations, hearings, and discussions on domestic satellite service for the U.S. The determination of the Commission, in its Final Report, modified the earlier AT&T/COMSAT restrictions; COMSAT was still required to form the separate corporate subsidiary in order to engage in DOMSAT ventures, and AT&T, its initial use of domestic satellites having been limited to its noncompetitive services, would now be permitted to openly compete in both competitive and noncompetitive markets after three years.[185] Here, the FCC had in mind the problem of preventing cross-subsidy and the identification of relevant costs, when AT&T combined its monopoly telephone services and competitive services in the same facilities.[186] The Final Report approved a proposal by COMSAT for joint and equal ownership in a restructured MCI Lockheed Satellite (in which no single owner, such as COMSAT, would control corporate management decisions) and the COMSAT servicing of AT&T with its separate subsidiary. AT&T divested itself of its COMSAT holdings forcing maximum mutual independence and COMSAT General, the COMSAT subsidiary, was formed.[187]

---

[185] Morrissey and Smith, "The FCC Activities," p. 753--"or upon a finding by the Commission that other specialized-services licensees have achieved substantial utilization of their satellite capacity, whichever is earlier." This was an attempt to minimize the effect of AT&T's monopoly on new competitors with no established or protected revenue bases. See also 38 FCC 2d 679.

[186] Bernard Strassburg, "New Technology and Old Institutions," Telecommunications (June, 1974), p. 25.

[187] Morrissey, "The FCC Activities," p. 756.

Although the Commission's decision on the Final Report was unanimous, the DOMSAT issue is viewed by others in many ways. Economist Alfred Kahn, the Chairman of the State of New York Public Service Commission says:

> In view of what I see as the Bell System's primary commitment to terrestrial communications modes, I can only feel that the introduction of competition, or entrusting the exploitation of communications satellite technology to some non-Bell chosen instrument are essential if this new technology is to be fully exploited. I find myself uncomfortable, however, when artificial restraints are placed on the competition by the Bell System itself, since it seems to me, in principle, satellites ought to be as fully integrated as technology requires with the total communications network; and putting limitations on AT&T's entry into that field, or on the uses that it can make of the satellites, it seems to me, may frustrate the achievement of that goal.[188]

The Common Carrier Bureaus Strassburg looked at the DOMSAT decision in this way:

> Basically we are embarking on a period of trial and error to determine what the benefits of domestic satellite will be and how the economics will spread.[189]

Later he noted:

---

[188] Letter from Dr. Alfred E. Kahn, Chairman, State of New York Public Service Commission, to Robert S. Magnant, June 16, 1976.

[189] "The DOMSAT Race is Now Wide Open," Business Week, September 22, 1973, p. 68.

> Our task as regulators is to see how effective we can be in maintaining the demarcation between two new areas of the communications market: the first is the telephone switched network service; the second is the competitive market. . .[190]

The 50-page order allowed the U.S. to join the U.S.S.R. and Canada as users of satellites dedicated to domestic communications. Burch, in a statement concurring with the FCC report, stated:

> Candidly, covering up the market-place and unfurling protective umbrellas are abhorent to me in principle--and the limitations we have imposed on AT&T's initial use of satellite technology clearly crossed the line in both respects. But we are not here dealing just with principle; we are in the grip of reality, and we must deal with it as best we can.[191]

Burch says:

> In retrospect, it seems like a fairly simple idea to prefer competition to monopoly, but at the time it was a rather momentous step. The members of the FCC, while I was there, were under considerable pressure to reject the "open skies" policy but, to their credit, were willing to think the unthinkable.[192]

c. Backlash to Competition

Almost immediately after the DOMSAT policy

---

[190]"Profile--Bernard Strassburg," Telephone Interconnect Journal (November, 1973), p. 32.

[191]Michael Sadofsky, ". . .This Proceeding is Terminated," Telecommunications (January, 1973), p. 21.

[192]Letter from Dean Burch, Washington, D.C., to Robert S. Magnant, May 12, 1976.

was finalized, the din of anticompetitive sentiment coming from the established carriers became more pronounced. Whether or not the DOMSAT policy of competition was the straw that broke the camel's back is hard to say. Perhaps satellite technology was seen by the established industry as having the greatest potential market impact. Prior to 1973, the common carriers had been fairly indifferent to the "so-called competition". But apparently times and attitudes were changing.

Shortly after Carterfone, Dr. Harry Trebing of Michigan State University referred to common carrier regulation as "The Silent Crisis" and called for a systems approach to future regulation, one that would recognize the coexistence of monopoly and competition. His praise of FCC policies was based on the Commission's efforts to accommodate the pressures for change by promoting greater flexibility in the domestic communications market structure. Additionally remarkable to him was the fact that these policies evolved in a regulatory framework which is usually accused of being reactionary and unresponsive to change.[193] But this is only one viewpoint of the situation. Senator Philip Hart had offered these perspectives of

---

[193]Harry M. Trebing, "Common Carrier Regulation--the Silent Crisis," Law and Contemporary Problems, 34 (Durham, North Carolina: Spring, 1969), pp. 325-329.

competition in the communications industries during this same period:

> Communications is one area--at least--where we should not get hung up on the perennial argument of whether antitrust exists to protect competition or competitors. In communications it must protect both. . .
>
> The new technologies should be moving us toward more, not less competition. . .
>
> Further, the regulatory agencies should encourage parallel systems of electronic transmission of written material. If success is achieved in this effort, then competition, not regulation, could determine rates.[194]

In 1970, Leland Johnson also raised questions of appropriate competitive boundaries between the traditional franchised monopoly and the competitive portions of the industry. He concluded that given the present-day and easily projected technological basis for the industry, the boundaries might reasonably be drawn between the so-called "switched" and "unswitched" portions of the nationwide network along lines previously suggested by President Johnson's Task Force.[195]

When Strassburg retired at the end of 1973, the communications industry suspected that the Commission's thrust to encourage competition with

---

[194] Philip A. Hart, "The Congressional Perspective of Competition in the Communications Industries," The Antitrust Bulletin, 13 (Fall, 1968), pp. 973-978.

[195] Johnson, "Technological Advance," p. 208.

AT&T might be blunted. Those hopes were quickly put down with the appointment of the new Bureau Chief, Walter R. Hinchman. Hinchman, originally an engineer from the Department of Commerce laboratories in Boulder, had come to the FCC via the competitive environments of the Johnson Task Force and the OTP. His initial concern was how AT&T itself competed and he directed his attention immediately to the elements of that issue (i.e., fair rates of return, cross-subsidization, economics of scale, etc.). He also took a dim view of the AT&T counter-offensive to stop or overturn the series of Bureau-supported FCC decisions permitting new competition.[196]

Counter-offensive? Yes, that is correct and directed right from the top. AT&T's Chairman John D. deButts is convinced that regulation and competition do not mix. In 1973, he called for a united front in the telephone industry to end "economic experiments" with competition. His philosophy? DeButts was quoted:

> My instinctive reaction to emerging competition matches that of our first president, who declared, 'We have established and organized the business, and we do not propose to have it taken from us.'[197]

---

[196]"New Chief of the FCC's Common Carrier Bureau Backs Competition in Telecommunications," Electronics (January 24, 1974), p. 31.

[197]"Who Will Supply the Office of the Future?", p. 49.

His message was contaminating and a campaign of unmatched sophistication was prepared by the AT&T headquarters to bypass the FCC. At a meeting of company presidents in November 1973, AT&T Vice President Paul M. Lund directed the Bell executives to hammer home to the public that "competition can be all bad for the telephone subscriber. . .Take the battle to politics. . ." he declared, ". . .to municipal offices, state houses, governor's mansions, meetings of civic clubs, engineering societies, services organizations. . .probing and maneuvering for concensus."[198]

What happened to the deButts, who in a 1970 speech said, "If others can make a contribution. . .it would seem unreasonable to deny them access,"[199] or the deButts, who later stated:

> . . .our experience of the consequences of the competition from the so-called specialized common carriers in the private line field has been relatively limited. Indeed, in a very real sense, it might be said that we have had no experience in such competition at all.[200]

DeButts justifies his basic change of mind in competition:

---

[198]Ibid., p. 49.

[199]Ibid.

[200]"Who Says Competition Isn't Good for the Market-place? Ma Bell, That's Who!" Telephone Interconnect Journal (October, 1973), p. 20.

> At first we had no experience with it. Now that we have had a couple of years of experience we know that it causes an increase in troubles. And if it continued, there would be an impact on rates.[201]

Economist Joseph Schumpeter once observed that no monopoly that acts like a monopoly can expect to survive for any length of time. He reached that conclusion not by looking at the economists' static model of monopoly behavior, but through a profound insight into the dynamics of our economic system. By looking at the long run, he saw not competition, but competitions--a whole myriad of forces that could undermine the unwary monopolist. And there is impressive historical evidence to support his contention.[202]

The traditional idea of telephone service has been replaced by the much broader concept of total communications. Total communications means getting the right information to the right place at the right time--and by information is meant either voice, data, or video service, carried by wire, cable, microwave, or satellite.[203]

---

[201]"Who Will Supply the Office of the Future?", p. 50.

[202]Paul H. Henson, Moderator, "Panel on 'Competition in telecommunications'," Proceedings of the Thirteenth Annual Iowa State Regulatory Conference (Ames, Iowa: Iowa State University, May 1974), p. 27. See also note 159, *supra*. The introduction of the telephone, which undermined message telegraph service, is probably the best example of this.

[203]Earle G. Bellamy, President of the United

It appears that AT&T has fallen victim to the forces of competition and technology, the impact of which it failed to anticipate. Nicholas Johnson believes that AT&T failed to conceptualize itself as in the 'communications' business rather than the telephone business. What was also to have a great impact, as Johnson saw it in 1969, was the fact that "the telephone company has failed to adequately anticipate and prepare for the present and future demand for communications service for computers."[204]

This set the attitudes for today's DOMSAT activity and provided the background for the entrance of the new kid on the block Satellite Business Systems.

. . . .

The multi-faceted world of the FCC has been shown through the example of the DOMSAT issue. Although every issue of Commission policy making is certainly not as dynamic as was that of the satellite, technology will most likely keep the Commission's world just as complex.

During the period of DOMSAT, the pace of technological change prompted the Commission to re-evaluate the natural monopoly thesis of the market. This applied not only to traditional areas where it

---

States Independent Telephone Association (USITA), quoted in <u>Public Utilities Fortnightly</u>, <u>81</u> (February 29, 1968), p. 47.

[204] Murphy, "Federal Regulatory Policy," p. 342.

was prevalent, but also to the new services that were being brought about by the rapid confluence of computers and communications. The agency, through a series of aggressive decisions of which DOMSAT was the last, injected the serum of competition into the telecommunications industry in an attempt to treat, if not cure, its regulatory ills.

Without a doubt, both the FCC and the Nixon Administration moved towards an active policy of competition. However, the political pressures that an Executive element like the OTP could promote do not seem to have impacted the Commission's decision significantly. The FCC appears to have been influenced much more by its own past policies, which had become more visible at the time of the OTP's formation.[205] The FCC's response supported competition once again. The primary criticism raised was that its DOMSAT policy was so long in coming. To some extent this delay can be justified by the complexity of the issue and the complications that were introduced by interrelated factors of technology, politics and market economics.

Burch and the Commission recognized that the deferral of certain aspects of the decision-making

---

[205]Trebing has reaching similar conclusions on this point. See Harry M. Trebing, "A Critique of Regulatory Accommodation to Change," Regulation in Further Perspective (Cambridge, Massachusetts: Ballinger Publishing Co., 1974), pp. 41-65.

process to the marketplace was both a good and necessary thing and "the way the Commission ought to operate." However, the international considerations of DOMSAT, although they were recognized, did not attain sufficient visibility in the proceedings to insure that today's global impact of DOMSAT would be recognized by the people of this Nation. There is no doubt that the Commission's objective in DOMSAT was to establish a regulatory framework which was intended to bring the benefits of advanced telecommunications technology to the public as soon as possible and that the Commission, in assuming its active role, was attempting to insure fair market competition. But space resources are global and more attention must still be given to international regulatory considerations and the role of the ITU in this area.

Jonathan Rose, Deputy Assistant Attorney General of the Justice Department's Antitrust Division has recently noted:

> [The] Above 890 [decision] was decided by the FCC without the intervention of other Federal agencies such as the Justice Department. Similarly, the FCC's Carterfone ruling. . .and the MCI-Specialized Carrier decisions. . . came about as the result of FCC initiations.[206]

The fact that the Nation's common carriers are a part

---

[206] Jonathan C. Rose, "Common Carrier Regulation and Antitrust Policy," Communications News (April, 1976), p. 16.

of the regulated industry only made the roles of competition and the Commission more important.

CHAPTER IV

DOMSAT POLICY TODAY

By 1973 the Commission's DOMSAT ruling was history and its role of regulator was more difficult than ever, primarily as a result of the dynamics of the evolving industry and its technology. There were five major contenders for this new market (or eight, depending on how the partnerships were counted) that the FCC had to regulate.

Reviews of the "DOMSAT game", with listings of all the players, have already been developed in John McDonald's The Game of Business and Michael Kinsley's Outer Space and Inner Sanctums [cited in the previous chapters] and will therefore not be presented here. Nevertheless, reviewing Satellite Business Systems (SBS), one of the game's newest players, provides one vehicle for looking at some of today's trends in technology and for examining the challenges that the FCC's overall competitive policy, which includes DOMSAT, is being subjected to today.

The ALOHA concept of data communications which is also described in this chapter, has provided an alternative transmission medium to the data communications designer. Although "pure ALOHA" is basically an undisciplined use of the capacity of a communications channel, it may, when used with satellites, be one method for extending shared information processing to the general public, from which a variety of beneficial services could be obtained.

SBS is the successor of MCIL Satellite, which was restructured (with COMSAT participation after the final DOMSAT decision) and renamed CML Satellite Corporation. When IBM announced intentions to acquire 55 percent of CML in July 1974 and get into a new business area, screams came from all directions; some factions believed that the move would stifle "competition"! Even the Federal Trade Commission warned of possible anticompetitive effects in a brief filed with the FCC. Actually, this announcement should not have come as too much of a surprise since IBM had been previously involved in a joint effort with COMSAT in 1967, testing the feasibility of using satellites for wideband data transmission between the U.S. and Europe.[1] The Commission turned down IBM's initial request in January, 1975 but qualified its decision by allowing that a joint venture could be approved if it included one or more additional partners, none of whom would own more than 49 percent nor less than 10 percent of

---

[1]Stuart L. Mathison and Philip M. Walker, Computers and Telecommunications: Issues in Public Policy (Englewood Cliffs, New Jersey: Prentice-Hall, Inc., 1970), p. 9.

the restructured CML.[2]

On September 27, 1975 Aetna Life Insurance announced that it was teaming up with IBM and COMSAT for a one-third interest in CML Satellite, with plans to build a satellite system largely for the transmission of data instead of voice communications and observers predicted that CML's revenues would reach a billion dollars in the early Eighties, provided it received Commission approval to proceed.[3]

Projected data transmission requirements show this market to be extremely promising; it is large and basically undeveloped. This and the IBM/SBS entry have added a sense of urgency to an AT&T-led campaign to minimize the effects of competition. The result is that proposed legislation is now before the Congress to reverse the FCC's competitive policies and petitions are now before the Commission to deny the SBS filing.

A. Trends in Technology

Some observers see SBS's proposed entry into the DOMSAT market to be in response to the increasing requirement for computer communications. Much of the technology of the DOMSAT systems operating or

---

[2] "Commission Decides on IBM/COMSAT General Joint Venture," Telecommunications Reports, Vol. 41, No. 4 (January 27, 1975), pp. 1-3.

[3] "Open Skies", Barron's (October 6, 1975), p. 9.

planned today is of an "early COMSAT/1960's" vintage, which does not meet the increasing requirement for high data rate communication systems by which computers can converse [the pessimists though see the pending entry as an IBM effort to dominate the data communications market and enlarge its computer markets]. In addition, there has been increasing interest in packet switching, intelligent networks, and distributive networks--terms which describe the concept of providing intelligibility to a communications network. Such a network comprises a number of geographically dispersed nodes or remote processors, interconnected by one or more communication channels.[4] Therefore, before looking at IBM's DOMSAT entrant, it is beneficial to review some current examples of computer network techniques, to look at the projected data traffic volumes for 1980 and consider the role that domestic satellite systems can play. To say that such ideas are simple in concept is not to minimize the difficulties of connection, control and programming involved, but these difficulties are being approached today in both obvious and ingenious ways.[5]

---

[4]John F. Buckley, "Network Node Criteria," Computer Design, 14 (December, 1975), p. 10.

[5]Stanley Winkler, "Computer Communication--the Quiet Revolution," Computer Communication--Impacts and Implications (ACM/ICCC, 1972), p. 30.

1. Data Projections

In recent years it has become commonplace for representatives of the computer industry and others to speak very convincingly of the enormous amounts of data that will soon be transmitted via communications lines throughout the U.S., quoting such authoritative sources as AT&T itself for their information. A 1970 study, performed at Stanford Research Institute for the National Aeronautics and Space Administration,[6] shed some light on this subject by addressing the potential demand for "information transfer" in a broad sense. That is, considering the fact that almost anything written, spoken, photographed or otherwise recorded can be transmitted electrically, what might be the maximum potential demand for all such information transfer activities if appropriate means were made available to implement them?

The study first assembled an extensive list of existing or potential information transfer services. These included electronic mail, remote library browsing, checkless society transactions, video-telephone service and so on, as well as conventional services such as telephone, telegraph and network television program distribution. Next, sample services were selected

---

[6] R. W. Hough et al., A Study of Trends in the Demand for Information Transfer (Menlo Park, California: Stanford Research Institute, 1970).

from the list on the basis either that something was known about the service or that it looked as though it might become important if it were to become an established service. An acceptable projection variable was then determined for each of the selected services.[7]

The final step was to convert all calls, messages, transactions and the like to a common denominator, bits (the basic unit of information theory used to describe message content) of information per year, by applying a conversion factor related to the transfer mode assumed for each type operation. The result was a statement of potential information transfer volume, expressed in bits, for each service. This information is illustrated in Figure 3.

In an editorial comment, Green and Lucky noted that this extensive analysis of the present and future demands for digital communications which SRI was still the most up-to-date information available.[8] A recent Department of Commerce, Office of Telecommunications Report has updated and added support to this statement.[9] The OT Report used

---

[7]Roger W. Hough, "Future Data Traffic Volume," Computer (September/October, 1970), pp. 6-12.

[8]Green and Lucky, Computer Communications, p. 4.

[9]R. L. Gallawa et al., Telecommunications Alternatives with Emphasis on Optical Waveguide Systems (Washington, D.C.: U.S. Department of Commerce, OT Report 75-72, October, 1975). Chapter II of this report provides an excellent general discussion

Source: Adapted from OT Report 75-72.

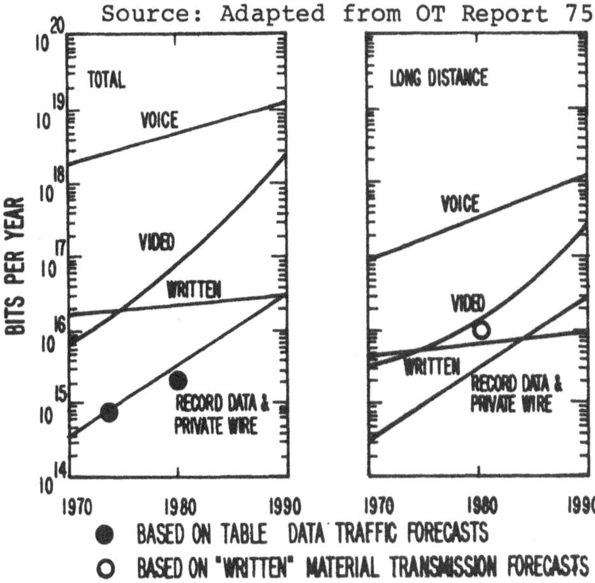

Figure 3. Projected Information Transfer Volume, 1970-1990.

estimates of growth rates for data transmission services, such as private wire, teletype (TWX and TELEX) and telegraph, that were independently derived from market research tables developed by Datran and contained in Computers and Telecommunications.[10] The data are shown in Table I. Gallawa et al. have additionally extracted data from other current Department of Commerce reports and used them also to further update the SRI predictions. Using electronic mail as the example, the estimated projected transmission

---

concerning the advantages of digital communications and the functions forcing the analog to digital evolution.

[10] Computers and Telecommunications, OECD Informantic Studies 3 (Paris: OECD, 1973), pp. 147-166.

TABLE I

DATA TRAFFIC FORECAST FOR THE UNITED STATES

	1970	1974	1980
Number of Transactions (billions)	14	50	250
Number of calls (billions)	3.7	12	32
Number of data terminals (thousands)	185	800	2,500
Number of termination points (thousands)	84	310	1,000

requirement is approximately 10^{16} (ten million billion) bits of information per year, which correlates with the SRI forecasts (as shown in Figure 3) and reaffirms the basic SRI predictions in order that they may continue to be considered valid for probable communications volume in coming years. The predictions that SBS has submitted to the FCC in support of its pending satellite venture, shed additional light on this area in terms of potential markets and are addressed later in this chapter.

2. Network Developments

There was considerable uncertainty before 1965 about incorporating geostationary communication satellites into the existing telecommunications plant. However, since then satellite channels have been used <u>almost exclusively</u> as replacements for cable or microwave channels. Only recently does it appear that the special capabilities of satellite channels are becoming more widely recognized. These capabilities are largely a result of the computer/communications research of Dr. Norman Abramson and his staff [which began in 1968] at the University of Hawaii and some of the earlier work done by the COMSAT Laboratories in data transmission.

a. SPADE

SPADE is a system developed for INTELSAT by the COMSAT Laboratories. SPADE, an acronym for Single-channel-per-carrier, Pulse code modulation, multiple-Access, Demand-assigned Equipment, was developed to utilize the direct multi-point distribution characteristics of satellite communications. SPADE is appropriate to this discussion since it links small processors (via satellite) which perform the signalling and switching functions for the system.[11]

Preassignment of satellite circuits, using full-time dedicated carriers, provides efficient system operation for links with heavy traffic. As the number of circuits per link is decreased, service becomes increasingly inefficient. One solution is to share a pool of satellite circuits among Earth stations in

[11]Eugene R. Cacciamani, Jr., "The SPADE System as Applied to Data Communications and Small Earth Station Operation," COMSAT Technical Review, Vol. 1, No. 1 (Fall, 1971), pp. 171-182.

common view of the satellite. In the SPADE system, a chosen satellite RF frequency band is divided on the basis of assigning a single voice channel per RF carrier, forming a pool of frequencies for circuits that can be accessed on demand by any station or terminal. The system uses a Demand Assignment Signalling and Switching (DASS) unit for self-assignment of channels based on continually updated channel allocation status data provided to a Common Signalling Channel (CSC).[12]

All signalling between SPADE terminals is conducted via this single "broadcast" channel. This feature plus a common signalling format allows additional terminals to be added, up to the system maximum, without requiring hardware or operational modifications to those terminals already in operation. A Signalling and Switching Processor (SSP) does the information analysis at each SPADE terminal. The common signalling channel is a time-shared data channel, relayed through the satellite, and shared by all SPADE terminals within the designated community of Earth terminals. Each operating terminal will transmit one short data burst of signalling information within each system time frame. These are received consecutively

---

[12] Andrew M. Werth, "SPADE: A PCM FDMA Demand Assignment System for Satellite Communications," Proc. Intelsat/IEEE Conf. on Digital Satellite Communications, 1969, pp. 51-68.

by the satellite and rebroadcast back to all terminals. Thus, each terminal will receive a data burst from all operating terminals during each system time frame. Channel performance parameters are shown in Table II.[13]

TABLE II

COMMON SIGNALLING CHANNEL
PERFORMANCE PARAMETERS

Transmission Bit Rate	---	128,000 bits/sec
Effective Signalling Bit Rate per Terminal	---	960 bits/sec
Bit Error Rate (Maximum)	---	less than 10^{-7}
System Frame Length	---	50 milliseconds
Total No. of Terminals	---	49

Data bursts include bits for carrier recovery, bit timing recovery and unique word synchronization as well as information bits and message error detection bits. One station in the system is designated as the reference station which transmits an additional burst which contains the uniquely identifiable sync word which denotes start-of-frame. Transmission is implemented using a two-phase PSK modem. A BCH Coder-Decoder

[13] G. Dill and N. Shimasaki, "Signalling and Switching for Demand Assignment Satellite Communications," Proc. Intelsat/IEEE Conf. Digital Satellite Communications, 1969, pp. 297-307.

is used to detect up to four errors in a 55 bit data block.[14]

The SSP was implemented using a small, high speed 16 bit parallel stored program computer. Since the functions performed are logical in nature and highly repetitive, this type unit is ideally suited for this purpose. Primary operational parameters of these computers are typically those identified in Table III. Typical cycle times achieved in existing SPADE terminals are 180 to 200 microseconds. Since the SPADE signalling channel uses a time division multiple access technique, each burst must contain a sufficient number of bits preceding the information and bit times between bursts to assure proper synchronization and no burst overlap (a 55 bit preamble and 7 bit times were assigned for synchronization and quard time respectively).[15]

b. ARPANET

The ARPANET, an experimental communications system of the Defense Department's Advanced Research Projects Agency (ARPA), has been extremely successful. Much has been written on this topic and it is not intended that this discussion will do any more than

---

[14]Werth, "SPADE: A PCM FDMA Demand Assignment System for Satellite Communications."

[15]Dill and Shimasaki, "Signalling and Switching for Demand Assignment Satellite Communications."

TABLE III

SIGNALLING AND SWITCHING PROCESSOR PARAMETERS

- A total of 8192 words of random access memory (minimum 16 bit words)
- A direct memory access channel (worst case access time--4 microseconds)
- Program protect capability
- Full cycle programmed memory access of less than 1 microsecond
- An automatic power failure and restart capability

---

identify the concept as a forerunner of subsequent developments. Design objectives for ARPANET were to develop and test computer-communications techniques and to ultimately benefit from the resulting resource sharing. A fully distributed switched message service was the configuration chosen for the system with the intent of obtaining greater reliability.[16]

A store and forward system must deal with the problems of routing, synchronization, error control and other related issues. To insulate computer centers from such problems and conversely the communications network from computer center problems, identical small processors were chosen to be located at each node of

[16] Lawrence Roberts and Barry Wessler, "The ARPA Network," Computer-Communication Networks (Englewood Cliffs, New Jersey: Prentice-Hall, 1973), pp. 485-500.

the network, interconnected to form a subnet by leased common-carrier circuits. These Interface Message Processors (called IMP's) connect the computer centers (called Hosts) together with the network, breaking the system design into two parts:

(1) protocol for utilization of the network by the Hosts

(2) implementation of the subnet.[17]

A consideration that later surfaced was how to make this resource available to users without Host facilities or terminal-oriented time-sharing systems. An IMP with flexible terminal handling capability (called TIP), was chosen as the solution, rather than incumbering a nearby Host with composing and translation problems, which an independent terminal user would impose.[18]

Message communications within the subnet are completely autonomous, the average transit time being less than half a second. Transmission is in the form of "packets", each of 1000 bits maximum. Fifteen facilities had been interconnected by February 1971 and the community of users numbered approximately

---

[17] F. E. Heart et al., "The Interface Message Processor for the ARPA Computer Network," AFIPS Conference Proceedings, 40 (1972), pp. 551-567.

[18] S. M. Ornstein et al., "The Terminal IMP for the SRPA Computer Network," AFIPS Conference Proceedings, 40 (1972), pp. 243-254.

2000.[19] By March 1973 the network had grown to 35 nodes. Thus the concept of packet switching for data/computer communications emerged. If one projects the growth of computer communication networks like the ARPANET to a worldwide situation, satellite communication systems become attractive for intercommunications between widespread geographic areas. The SPADE system [or the TDMA demand-access system MAT-1] permits demand assignment of satellite capacity only on circuit-switched basis for minutes rather than in small blocks of data or "packets" to a variety of locations. Several techniques have been proposed and developed as will be shown.[20] Demand assignment techniques in general enhance the usefulness of satellite communications networks. They provide advantages where channel requirements are few and for overflow traffic for preassigned trunks. Power and bandwidth efficiencies are also increased.[21] Demand assignment in smaller elements or "packets" naturally becomes one of the next logical steps.

---

[19] Roberts and Wessler, "The ARPA Network." Figure 4 shows the ARPANET one year later.

[20] Lawrence G. Roberts, "Dynamic Allocation of Satellite Capacity Through Packet Reservations," National Computer Conference (1973), pp. 711-716.

[21] J. G. Puente and A. M. Werth, "Demand-Assignment Service for the INTELSAT Global Network," IEEE Spectrum (January, 1971), pp. 59-69.

Source: Roberts and Wessler, "The ARPA Network"

Figure 4. ARPANET-March 1972

c. <u>Packet Economics</u>

The predominant force for the past 20 years in both computer and communications architecture has been the rapid decrease in computer hardware costs. Packet switching is strongly dependent on the computing costs, since computation is used for dynamic bandwidth allocation, transmission error correcting and reliability through alternative routing. It is primarily because the cost of computing has been decreasing far more rapidly than the price of communications that such concepts have evolved.

The ARPANET in its present configuration provides a factor of ten or more in cost advantage over conventional circuit-switched systems. Available data can provide some idea of the economics that can be anticipated in transmitting data via a packet-type of service. During the month of January 1973 approximately 45 million packets were transmitted by the ARPANET, while its capacity (based on an eight hour day) was about 300 million packets per month.[22]

In March 1973, several contractors were making substantial use of the ARPANET for a majority of their computing resources. An accounting of the utilization at that time and the estimated costs for the network usage and comparable purchased time was developed (see Table IV). Over $2 million a year was being spent on computing resources assessed through the network, resources that would have cost $6 million a year if the network had not existed.

In general, the annual network costs (including the amortization of the message processors) were more than offset by the savings on resources but the significant point is that the network was only using twenty percent of its capacity. Further analysis demonstrated that for a moderate investment in communications, a computing cost reduction factor of three can be achieved by resource sharing in this manner.[23]

---

[22] Norman Abramson, "Packet Switching with Satellites," National Computer Conference (1973), pp. 695-702.

[23] Lawrence G. Roberts, "Data by the Packet," IEEE Spectrum (February, 1974), pp. 46-51. See also Roy Brunn, "Packet Switching and Satellite Technology," Infosystems, 21 (August, 1974), pp. 33-35.

TABLE IV

COMPUTER RESOURCE USAGE WITHIN ARPANET

User Organization/Activity	Remote Usage (thousands)	Projected Cost for Replacement (thousands)
U. of Illinois--Parallel processing	$ 360	$1100
NASA Ames--Air foil design	328	570
Rand--Numerical modeling	210	650
Mass. Computer--ILLIAC IV compiler	151	470
Lawrence Livermore--ILLIAC	94	370
Stanford U.--Artificial intelligence	91	180
R.A.D.C.--Text manipulation	81	450
ARPA--On-line management	77	370
Seismic Analysis--Data processing	76	300
Mitre Corp.--File network research	60	240
NBS--Network research	58	200
B.B. & N. Inc.--TENEX support	55	80
Xerox Palo Alto--Computer research	47	100
USC Image Lab--Picture processing	35	70
UCLA--Network measurement	28	90
Systems Control--Signal processing	23	70
U. of C., Santa Barbara--Network research	22	70
U.S. Air Force--ARPANET management	17	60
Institute for Future--Teleconferencing	13	40
Miscellaneous--Computer research	192	580
TOTAL	$2018	$6060

Source: Roberts, "Data by the Packet", IEEE Spectrum, February 1974.

Annual remote computer usage costs are based on March 1973 data.

d. Line Control Procedures

Rules have been evolving for the efficient use of data transmission facilities since J.M.E. Baudot developed his famous five-unit code more than fifty years ago. In early code transmission, "start" and "stop" bits were used to separate characters and to synchronize the receiving station with the transmitting station. Synchronous transmission schemes now permit more information to be passed over a circuit per unit time because no time is required or lost for the insertion of these signal elements. This discussion only presents Binary Synchronous Communications (BSC) as an example of a data link control (DLC) procedure that can provide for synchronous transmission and the attachment of multiple terminal devices to a transmission medium. BSC (or Bisync) enables the adapter/interface equipment to control data flow, maintains the synchronization between devices, and offers code flexibility and also transmission efficiency. It generally conforms to standard line control techniques of any system that has been designed to deal with the line errors, addressing, polling and multiplex problems associated with having an on-line system with multiple stations.[24]

Communication connections can be simplex

---

[24] IBM, Introduction to Teleprocessing--General Information, GC20-8095-02 (March, 1973), p. 30.

(unidirectional), half duplex (one way at a time but alternating directions) or full duplex (both ways simultaneously). While the purpose of a DLC is to acquire and maintain synchronization between separate machines, there is usually a higher order user of the communications facility that the DLC ultimately serves. The DLC must act as a conduit for some of the transmitted data, the contents of which should not affect or be affected by the DLC.[25]

BSC transmission is limited to half-duplex over 2-wire or 4-wire private leased or dialed/switched communications channels. All data must be synchronously timed and typically EBCDIC (Extended Binary Coded Decimal Interchange Code) or ASCII (American Standard Code for Information Interchange) is used although an IBM 6-bit Transcode can also be used. The codes comprising the primary control set with BSC are:

> Synchronizing Character (SYN)--is used to establish synchronization. Three SYN characters must precede, and one SYN character must follow, each continuous transmission.
>
> Enquiry (ENQ)--Depending on the state of the transmitting or receiving devices, this is multiple-purpose: solicit the remote device status, request retransmission of a response, and/or indicate an I/O error when transmitting.
>
> Start of Text (STX)--precedes a sequence or block of data characters.

---

[25] J. P. Gray, "Line Control Procedures," *Proceedings of the IEEE*, *60* (November, 1972), pp. 1301-1303.

Intermediate Block Check (ITB or US or IUS)--indicates the end of a record in a multiple-record block.

End of Transmission (ETB)--ends each data block within a multiple-block transmission.

End of Text (ETX)--is used in lieu of ETB for the last data block in a transmission.

Data Link Escape (DLE)--when preceding any character, this alters the meaning of that character; DLE plus any other character can be used to create a new control function.

Positive Acknowledgment (ACK Ø or ACK 1)--are actually character sequences: DLE followed by the numeric Ø or 1 defines ACK Ø or ACK 1, respectively. ACK Ø is used to provide a positive response to all even data blocks; ACK 1 is a positive response to all odd data blocks. The first block is considered odd, the second even, the third odd, and so on.

Negative Acknowledgment (NAK)--is a negative response to a data block received in error; it is also used to convey not-ready-to-receive status to the transmitting device.

End of Transmission (EOT)--terminates the correct transmission.

Block Check Character (bcc)--performs longitudinal record check.

Peripheral operations available with BSC include multipoint line control, auto-answer and EBCDIC transparency. Multipoint line control allows a number of terminals to operate with a host over a multipoint communications line. A polling or selection operation is initiated when the host transmits a 3-character identification sequence ending in an ENQ. Auto-answer enables the terminal to automatically answer incoming calls. Code transparency allows all possible bit combinations to be used as data. Code sensitivity that restricts special characters limits the power of the code being used.[26]

3. The ALOHA Concept

Prior to the ALOHANET, conventional methods of remote access to a large information processing system were limited to wire communications--either private leased or dial-up connections. In some situations, they were adequate; in others, the limitations imposed by wire restricted the usefulness of remote access computing. The use of satellite systems have only expanded the possibilities of this concept. The goal of the ALOHA system was to provide another alternative for the systems designer and to determine under what circumstances radio communications were preferable to wire for data transmission.

The general availability of wire systems is one of the obvious reasons for its widespread use in present day computer-communications systems (although in many parts of the world reliable high quality wire networks are not available!). One must remember, though, that the technology that spawned the voice/wire networks was one designed for analog signals. This in itself limits the digital/data transmission capabilities

---

[26]John E. Buckley, "IBM Protocols-Part 1: BSC," Computer Design, 14 (January, 1975), pp. 12, 14, 18.

possible and presents a number of drawbacks in the transmission of binary data.

Data transmitted in a time-shared computer system comes in a sequence of bursts with extremely long periods of silence between the bursts. If several consoles can be placed in close proximity to each other, multiplexing and data concentration may alleviate this difficulty. When this is not feasible, the user may find that his major costs arise from communication rather than computation and that his communication system is being used at less than one percent of its capacity.[27]

a. Pure ALOHA

When the designer of a computer-communication system is freed from the constraints imposed by the use of common carrier communications, a number of new possibilities present themselves. The ALOHANET uses a radio link. The existing computer-communications network uses two 24,000 bit-per-second channels in the UHF band. The system employs message switching techniques similar to the ARPANET with a novel form of random-access radio channel multiplexing.

The central computer of the ALOHANET, an IBM 370/158 (an IBM 360/65 prior to December 1974), is linked to the radio channels by a small interface

[27]Norman Abramson, "The ALOHA System--Another Alternative for Computer Communications," AFIPS, 37 (1970), pp. 281-285. See Figure 5.

Source: Abramson, "The ALOHA System"

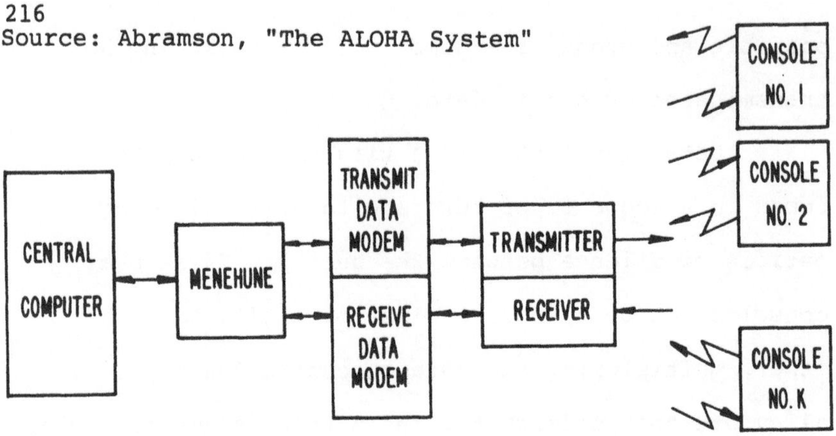

Figure 5. The ALOHA System

computer (whose design is based on the ARPANET IMP) that has been dubbed MENEHUNE, a legendary Hawaiian elf. The MENEHUNE is an HP 2115A with a 16-bit word size, a 1 microsecond cycle time and 8000 words of core storage.[28]

The ALOHANET has been assigned two 100KHz channels (407.350 MHz and 413.475 MHz), one for data from the MENEHUNE and the other for data from the remote consoles. From the MENEHUNE to the consoles is no problem since the MENEHUNE can control the transmitter and buffer messages for the different consoles. The queues can be ordered according to any given priority scheme and then transmitted sequentially. The system presently operates on a first-in, first-out

[28] R. Binder et al., "ALOHA Packet Braodcasting--A Retrospect," National Computer Conference (1975), pp. 203-215.

basis.

Messages from the consoles are not as easily multiplexed. If standard orthogonal multiplex techniques are employed, the channel would have to be divided into a large number of low-speed channels, each assigned to a particular console, whether it was active or not. Such schemes lead to the same sort of inefficiencies found in wire. The problem could even be partly solved by a system of central control or by a variety of polling techniques. However, ALOHANET uses packets in a common high-speed data channel.[29]

b. Random Packets

The random access method employed by the ALOHA system is based on the use of cyclic error detecting codes. Packets consist of a fixed length of 704 bits (80 8-bit characters plus 32 identification and control bits and 32 parity bits for error detecting) and are transmitted in 30 milliseconds at 9600 bps). Each subscriber transmits packets to the MENEHUNE over the single high data rate channel in a completely unsynchronized (user to user) manner. The transmitting console waits a given amount of time for acknowledgment. Only if a packet is received without error by the

---

[29]Norman Abramson, "The ALOHA System," Computer Communications Networks (Englewood Cliffs, New Jersey: Prentice-Hall, Inc., 1973).

MENEHUNE does an acknowledgment occur. If none is received, the packet is retransmitted.[30] Since all users access the same channel in 30 millisecond bursts, each automatically multiplexes their data into that channel at the time it's transmitted. The multiplexing is accomplished between the transmitting antenna at each station and the receiving antenna at the central station. One might say that "the medium is the multiplexor."[31]

It is assumed that all error patterns are equally likely. Thus the probability that a given error pattern will not be detected by the error code is 2^{-32} or approximately 10^{-9}. A transmitted packet can be received incorrectly because of two different types of errors: (1) random noise errors and (2) errors caused by interference with a packet transmitted by another console. The former is not considered a serious problem. The interference error is of importance only when a large number of users try to use the single channel and will limit the number of users and amount of data which can be transmitted.[32] Further analysis of ALOHA and its protocol considerations can be found

---

[30]Abramson, "The ALOHA System--Another Alternative for Computer Communications."

[31]Abramson, "Packet Switching with Satellites."

[32]Abramson, "The ALOHA System--Another Alternative for Computer Communications."

in Appendix C.

c. Satellite Channels

It was previously believed that a satellite system would be of no earthly use [a pun of former FCC Chairman Rosel Hyde] if it could not be connected with terrestrial facilities such as the telephone system. The following is an excellent example of how space technology has done just this and how it has expanded upon the scope of an existing application. A 50 kilobit INTELSAT channel is now being used to link the ALOHA system to the ARPANET and it could easily link additional nodes into the ARPANET at each of the 16 earth stations that have access to the Pacific satellite. For almost a century circuit switching has dominated the design of communications networks. With the higher speed and lower cost of modern computers, packet communications are competitive.

Rather than using satellite channels only as circuit paths, the broadcasting nature of satellites offers two additional unique properties to packet switching. First, not only the broadcasting nature, but the bilateral broadcasting nature of satellites offers certain advantages over normal point-to-point systems. In other words, consider the use of a single channel (like ALOHA) for broadcasting into by transmitters at all ground stations rather than the normal single-transmitter idea of broadcasting. The second property is that of information feedback. With a satellite system each ground station has the capability of transmitting data packets via the satellite addressed to any other station, but all ground stations receive each packet, including the ground station that transmitted it, approximately one quarter second later. With pure ALOHA it was necessary to provide packet interference information to the sender in the form of positive acknowledgment. But now, since each sender can listen to his own packet being retransmitted by the satellite, such information is not necessary. Thus more efficient use of negative acknowledgment schemes in conjunction with packet numbering is feasible for such a system.[33]

Cost trends are rapidly changing with the introduction of domestic satellite communications. If cost trends follow initial international estimates (see Table V), the projection of cost performance for domestic satellites will be very close to that of computers. The SBS system will take advantage of high frequency (12-14 GHz) satellite technology which will permit direct interconnection to the satellite with very small Earth stations and eliminate the lower limit of communications costs presently imposed by terrestrial interconnection.

---

[33]Abramson, "Packet Switching with Satellites."

TABLE V

COST ESTIMATES FOR INTELSAT COMMUNICATION SATELLITES

INTEL-SAT	Usage Year	Number of Circuits	Life-time years	Total Cost	Cost per Circuit per Year
I	1965-67	240	1.5	$ 8.2M	$22,800
II	1967-68	240	3	$ 8.1M	$11,300
III	1968-71	1,200	5	$10.5M	$ 1,800
IV	1971-78	6,000	7	$26.0M	$ 600
V*	1978-85	100,000	10	$28.5M	$ 30

*Estimated Source: Roberts, "Data by the Packet"

Because ALOHA channels operate in a burst mode with a duty cycle significantly less than one, the average power of the channel can be well below the peak power, just as the average data rate is below the peak data rate. In initial studies, throughput analysis was only compared to conventional point-to-point channels of the same peak power. More recent analyses have compared channels of the same average power.[34] Such analysis is of interest in the case of a satellite information system employing thousands of small Earth stations. The fundamental limitation in the down link is the average power available in the

[34] Norman Abramson, "Satellite Packet Broadcasting to Very Small Earth Stations," DCEC Rept. TN-38-75 (Reston, Virginia: September, 1975), 22 pp.

transponder. The results show that in the limit of large numbers of small Earth stations, ALOHA throughput approaches 100 percent of the point-to-point capacity. Multiple access and complete connectivity can be obtained at no sacrifice in average throughput. Higher power satellites (higher peak power while average power is kept constant) permits the small earth stations to use smaller antennas and simpler receivers and modems than is necessary for conventional systems.

Kleinrock and Lam did a model analysis of the performance of a slotted satellite system for packet switching with traffic from many small users (Model I) and for background traffic with one large user (Model II). In the case of Model I, the limiting throughput of the channel [1/exp] can be approached fairly closely without an excessive retransmission delay. With Model II, it was shown that it is possible to increase throughput rather significantly.[35] A single large user can transmit data at a significant percentage of the total channel data rate, thus allowing use of the channel at rates well above the limit of 37 percent obtained when all users have the same message rate. This capability is important for

---

[35] Leonard Kleinrock and Simon S. Lam, "Packet Switching in a Slotted Satellite Channel," Proceedings, National Computer Conference (1973), pp. 703-710.

a computer network consisting of many interactive terminal users and a small number of users who send large but infrequent files over the channel.[36]

It is worthwhile noting that another scheme is currently being investigated for packet-switching systems in which the propagation delay is small compared to the slot time. In such systems it may be advantageous for a user to "listen before transmitting" in order to determine channel use. Such systems are referred to as "carrier-sense" systems. For satellite communications this small delay/slot time ratio may be found when the capacity of the channel is rather small (approximately 1200 bps for purposes of this discussion), but it should be recognized that this condition is directly related to packet size. In a ground radio environment, a 50 kilobit channel with packets on the order of 100 or 1000 bits lend themselves nicely to carrier-sense techniques. All these techniques have complexity/performance trade-offs.[37]

A novel proposal considers the use of a single transponder in a domestic satellite system to provide a public-packet switched data communication service. One INTELSAT IV transponder could easily provide a

---

[36] Abramson, "Packet Switching with Satellites."

[37] Kleinrock and Lam, "Packet Switching in a Slotted Satellite Channel."

data rate of 10 Mbps (the proposed SBS system is to carry 41 Mbps per transponder). Assuming 100 Earth stations, each with small communications controllers, and burst data rates of 10 megabits, the capacity of such a system (in terms of interactive alphanumeric terminals) would be about 100,000 active users. Another interesting point is that the average power in a satellite transponder limits a channel. However, the satellite will only transmit power when it is relaying a packet so when its operation is in a lightly loaded condition, its duty cycle will be small and the average power used low.[38]

In February 1975, COMSAT filed with the FCC for authority to install and operate satellite interface message processor (SIMP) equipment at its Etam, West Virginia Earth station (on a two-year experimental basis in conjunction with single-channel-per-carrier equipment) to provide satellite packet switching to the ARPA Network. Service would initially be between the United States and the United Kingdom.[39] Information on the results to date indicates that the initial strides made in developing this service

---

[38] Abramson, "Packet Switching with Satellites."

[39] "COMSAT Proposes Two-Year Test of 'Packet' Switching," <u>Telecommunications Reports</u>, <u>41</u>, No. 6 (February 10, 1975), p. 36.

have been highly successful.[40]

As an aside, it is interesting to note that the question of "actual" packet switching economies becomes very complex when satellites are used as the transmission medium. In terrestrial systems, the value-added carriers (like Telenet Communication, who offers packet-switching services by means of a "commercial ARPANET") take advantage of the small incremental costs that are available when traffic loading is increased on trunks that normally carry intermediate levels of traffic. Routing information (overhead) is added to the basic message to be transmitted and, although actual throughput is decreased, more efficient use of the trunking capacity is obtained and the cost of service to the user becomes more economical because of decreasing incremental costs. When trunking facilities are provided by satellite, because of the "bird's" inherent multipoint distribution characteristic the economic picture becomes quite different and there are a variety of opinions that exist over what the "actual economies" are.

Strangely enough, despite these mixed opinions, an international standard for packet network protocol was arrived at this spring in Geneva.[41] Although the

---

[40]Personal Interview with Philip M. Walker, Vice President and General Counsel, Telenet Communications Corporation, July 7, 1976.

[41]This has been accomplished rapidly because

standard, recommended by the Consultative Committee on International Telephone and Telegraph (CCITT), must still be approved at a plenary meeting this fall, this appears to pose no problem.[42] Recommendation X25, as it is known, supports two basic types of transmission--permanent and switched virtual calls. Both require a duplex point-to-point synchronous link between the sender and receiver and his network node. Telenet has already announced that it will implement the protocol. IBM has indicated that it also plans to support it but has not yet said how long it will take to make the necessary changes in their existing address control scheme--Systems Network Architecture (SNA).[43]

B. The IBM Entry--SBS

Armed with the concepts of computer networks, packet switching and satellite channels, a discussion of the impact of the SBS filing should now be more

---

basically no international standard of communication protocol has previously existed. This is a result of the fact that in the past most data traffic has been passed over private-line circuits or that routing or header information has been provided by the dial network. [I must thank Mr. Thijs de Haas of the Department of Commerce for helping me to understand this complex subject.]

[42]"Protocol for Packet Networks: The Question is Implementation," Datamation (May, 1976), pp. 187-188.

[43]Ibid.

meaningful to the reader. A detailed technical analysis of satellite system is not intended but rather an appreciation for the driving force of technology and for the issues that have been raised regarding the DOMSAT policy as applied to the IBM entry.

Upon reading the Commission's thirty-six page Memorandum Opinion and Order on Docket No. 20221, the restructuring of the CML Satellite Corporation, there can be no doubt of the profundity of the basic DOMSAT ruling. The factors relating DOMSAT to interconnect, competition, specialized carriers and Computer Inquiry considerations become more obvious. In December 1975, SBS became the latest applicant for a domestic satellite system and the first to file with the FCC for a system in the 12 and 14 GHz frequency bands. This filing contained the first published description of the SBS system design.[44]

Within ninety days, the AT&T-sponsored "Consumer Communications Reform Act of 1976" was introduced into both houses of Congress. It was not just a reaction to the filing; it was a reaction to the basic idea of the IBM entry, a spectre that had been raised almost

---

[44]For a summary of the system description of the SBS filing, see C. Kittiver and F. R. Zitzmann, "The SBS System--An Innovative Domestic Satellite System for Private-Line Networks," presented at the AIAA/CASI Sixth Communication Satellite Systems Conference (Montreal, Canada: April 5, 1976).

two years earlier.

1. A Joint Proposal

In August 1972 Joseph H. McConnell, Chairman of COMSAT, announced that COMSAT had signed a Memorandum of Understanding with MCI Lockheed Satellite Corporation to develop a jointly owned multipurpose DOMSAT system, contingent upon the FCC's approval of such a joint proposal.[45] After the Commission's Final Report approved the COMSAT participation, CML Satellite, an attempt to combine COMSAT's technical expertise with MCIL's marketing ability, livened up the competition. CML proposed to use larger and more sophisticated satellites than any of the other contenders[46] and the newer untried frequencies (12 and 14 GHz) which would permit satellite signals to be beamed, without interference, directly to rooftop satellite terminals, avoiding "the costly dependency on interconnections." As Kenneth Crandell, development and planning director put it, "Our motto is: 'don't fight the interconnection problem, avoid it.'"[47] But CML had other problems. MCI and Lockheed, both

[45] "COMSAT at the Tenth Anniversary of the Satellite Act," Public Utilities Fortnightly, 90 (September 28, 1972), p. 40.

[46] "The 'DOMSAT' Race is Now Wide Open," Business Week, September 22, 1973, p. 70.

[47] "DOMSAT a Year Later: Tight Race," Electronics (June 21, 1973), p. 72.

of which were having financial difficulties, were looking for buyers for their interests--being unable to make the necessary large capital investments needed.[48]

In a move that surprised both the computer and communications industries, IBM announced on July 3, 1974, that it had agreed to take over fifty-five percent of CML, joining with COMSAT in buying up the Lockheed and MCI interests for a total of $5 million. With its market for computers and data processing systems increasingly dependent on communications links, IBM was prepared to apply its financial and technological muscle to the task of building a new kind of domestic telecommunications network.[49] It thus became the first major communications user with no previous stake in the industry to invest in satellite communications.[50] "Any way you look at it," noted Commission economist Dr. Manley R. Irwin, "IBM's entry has to be considered a major fork in the road."[51]

---

[48]"What Launched IBM into Satellites," Business Week, July 13, 1974, pp. 24-25. A recall of seven Earth station applications in February 1974 had been described as "housekeeping" actions by CML.

[49]Ibid., p. 25.

[50]Michael E. Kinsley, Outer Space and Inner Sanctums (New York: John Wiley and Sons, 1976), p. 191.

[51]"Who Will Supply the Office of the Future?" Business Week, July 27, 1974, p. 45.

In a joint statement, IBM's chairman Frank T. Cary and COMSAT's president Joseph V. Charyk said, "By joining technical capabilities and experience, we believe we can make a substantial contribution to the development of satellite communications."[52] Sebastian A. Lasher of the White House Office of Telecommunications Policy (and now a consultant to FCC Commissioner Washburn) had guessed right when he told an International Communications Association Conference in January, 1974:

> We can recall the Kingsburg Commitment in 1913, the establishment of the FCC in 1934, and the ATT-Western Electric consent decree in 1956. It seems to me that the time is ripe again in the mid 1970's for this '20-year itch' to manifest itself again. The next five years may be critical for this is a period where the opportunities of change seem to outweigh the risks of change in telecommunications.[53]

Although Cary and Charyk agreed, there were others that did not quite see it that way.

Jack Biddle, Executive Director of the Computer Industry Association, complained that if the IBM-COMSAT joint venture were approved it would enable IBM to monopolize the telecommunications business "in the same way it has monopolized the computer business." He

---

[52] "IBM to Invest $3.2M in CML," Electronic News (July 8, 1974), p. 26.

[53] "Who Will Supply the Office of the Future?", p. 50.

further contended that rather than competing, "AT&T and IBM will simply divide up the market."[54]

Testifying before Senator Hart's Antitrust and Monopoly Subcommittee, Royden C. Sanders, Jr., president of Sanders Associates, Inc. noted:

> Domestic satellites are the key to low-cost communications. A properly implemented switched satellite system providing neutral or transparent interconnections to all users, with no built-in bias toward one equipment supplier, would be very beneficial to the American public and the industries which supply it.
>
> . . .technology exists today that would enable IBM to simply bypass much of the local telephone plants now operated in urban areas by AT&T for business communications by the use of a very small rooftop antenna that communicates directly with the satellite.[55]

However, he warned that those viewing IBM's entry as a panacea for the AT&T monopoly problem were making a catastrophic mistake for IBM's dominance of all elements of a distributed system would make it impossible for meaningful competition to arise.[56] He recommended that the IBM/CML acquisition be denied

---

[54]"IBM Too Big for a Satellite Venture?", Datamation (August, 1974), p. 99.

[55]U.S. Congress, Senate, Subcommittee on Antitrust and Monopoly, Hearings on the Industrial Reorganization Act, S. 1167, Part 7--the Computer Industry, July 23-26, 1974, pp. 5436-5448. Emphasis added. This issue was raised again in September, 1974 by the Computer Industry Association.

[56]Ibid., p. 5438.

because of the potential for market foreclosure and anticompetitive maneuvering.[57] MCI and Lockheed were ready to sell but the Commission's blessing would have to be obtained first. A petition was filed on July 15, 1974 with the Commission and Docket No. 20221 was opened.[58]

2. The CML Decision

While IBM was saying little about its proposed system, it did note that it would "be different from the first generation systems proposed by the other entrants." It also acknowledged that it would be "a digital satellite transmission system for integrated voice, image and data service, including high-speed data transmission on a multipoint-network basis."[59] In support of the joint petition, CML, IBM, COMSAT and COMSAT General claimed that the financial resources of COMSAT General and IBM would provide assurance in sustaining CML through both the high risk/high cost development stage and later periods when continued long-range viability "will depend on its ability to meet the competition of existing broad-based

---

[57]Ibid., p. 5447.

[58]See Memorandum Opinion and Order, Petition for Approval of Changes in Corporate Structure of CML Satellite Corp., Docket No. 20221, 51 FCC 2d 14.

[59]"IBM Plan: Individual Earth Stations," Electronics (November 14, 1974), p. 67.

communications companies."[60]

Surprisingly AT&T was playing it markedly cool, in sharp contrast to its determined opposition to terrestrial competition. It noted that it had no objection to the proposed change in ownership in CML but reserved the right to address the "public interest questions" it expected to be in CML's formal applications.[61]

Of the 17 parties who provided comments in response to the public notice issued on the joint petition filing, most noteable were the Federal Trade Commission (FTC) and the Computer Industry Association (CIA). The Justice Department, expressing interest in the proposed acquisition, asked the Commission to delay any decision until the question of any possible antitrust violations was resolved.[62] The FTC stated that the IBM/COMSAT petition:

> could have serious anticompetitive effects on the data processing market and sub-markets, and on the integrated business systems market, particularly as they are enhanced by satellite communications.[63]

---

[60] 51 FCC 2d 23.

[61] "IBM Plan," p. 67.

[62] "Heard on the Street", The Wall Street Journal (October 25, 1974), 27:3.

[63] "Trade Commission Opposing Joint Venture," Telecommunications Reports, 40, No. 39 (September 30, 1974), p. 17.

They urged "a full and complete factual and legal inquiry into the matter."[64]

The Computer Industry Association is probably the most fearful and most outspoken of all IBM critics. In the September 1 issue of the CIA newsletter <u>On Line</u>, the fear of a CML "lock-in" under which users would be persuaded to rent everything--central computer systems, remote systems, intelligent terminals and even data communications circuits--from a single supplier at least indirectly was strongly evident.[65] It alleged that the IBM "systems" marketing approach would preclude the use of other than IBM equipment and that since no other domestic satellite entrant would be able to provide such "bundled" service, a large segment of the competitive DOMSAT market would be dominated by IBM/COMSAT.[66]

a. <u>SDLC</u>

The CML DOMSAT was seen by the CIA as the missing link that would integrate IBM's computer systems and IBM's "Carnation" private exchange

---

[64]"FTC Slaps IBM-COMSAT Venture Plan," <u>Electronic News</u> (September 30, 1974), p. 1.

[65]"The Ultimate Lock-In?", <u>Modern Data</u> (June, 1975), p. 26.

[66]51 FCC 2d 21.

equipment (PABX-Model 3750) that was being made and successfully marketed by IBM in France. Another IBM development was Synchronous Data Link Control (SDLC), a communications line protocol that provided for increased transmission efficiency. This was seen as a means by which the proposed DOMSAT system could be designed with a built-in bias for IBM to deal "a blow to competition in the industry."[67]

By comparing SDLC with its forerunner, BSC, and considering the objectives of its development (see Table VI) such myths can be discarded. The "frame", as shown in Figure 6, is functionally analogous to the "block" terminology used with BSC. SDLC is a data link control for serial-by-bit synchronous transmission between buffered stations on a data transmission link using centralized control. It is bit-oriented and any receiving error invalidates the entire transmission.

Generally the frame comprises a maximum of six field positions; of these two are called flags (F) and they begin and end each frame (with the bit configuration 01111110). The (A) field is the address of the secondary station and is always eight consecutive bits. The primary station has the centralized control. The (C) field, also eight bits, contains

---

[67]"The Ultimate Lock-In?", p. 28.

TABLE VI

SUMMARY OF MAJOR REQUIREMENTS

- Control of the data link should remain with a central station to facilitate loop operation as well as to simplify error recovery procedures.

- Transmission block length should be independent of message or record length.

- The DLC should allow straightforward transmission of any message. It should not be possible for any bit pattern within a transmission to be mistakenly construed as a control character.

- The DLC should not impose unnecessary line turnarounds.

- Operation on systems with a wide range of line speeds and propagation delays should be possible.

- The basic DLC structure should be equally applicable to half-duplex, duplex, reverse channel, hub, or loop link configurations.

- A simple form of the DLC should be provided for use with simple stations, with more complex forms available for more intelligent stations.

- The amount of redundancy for checking purposes should be variable.

- A common discipline should be used for recovery from detected errors in the communications link.

- The addressing and control structures of the DLC should be open-ended.

- The DLC should minimize the down time of all elements of the system.

- The DLC should provide a means of obtaining status information from the various system elements.

Source: R. A. Donnan and J. R. Kersey, "Synchronous Data Link Control: A Perspective," IBM Systems Journal, Vol. 13, No. 2. (1974).

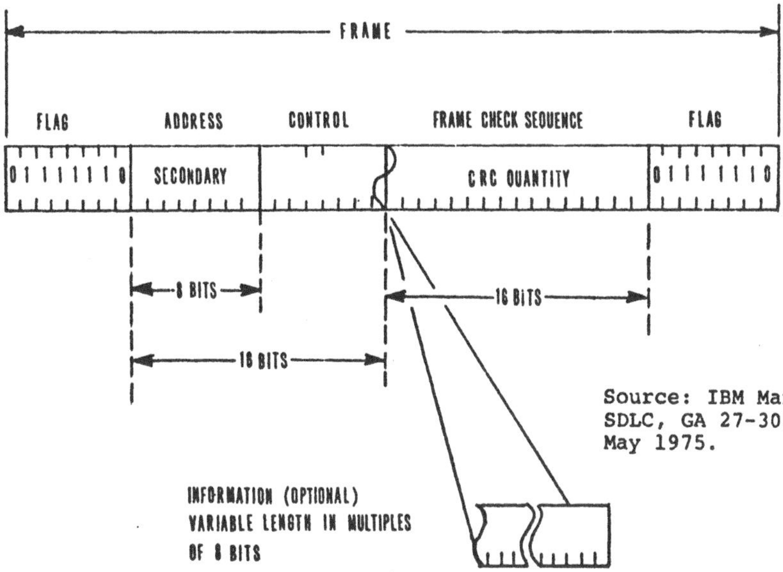

Source: IBM Manual SDLC, GA 27-3093-1 May 1975.

Figure 6. The SDLC Frame

control information. It may perform either a command or a responsive function. The (I) field is the position of the actual information to be transmitted. Physical limitations may be imposed on this field by a terminal but protocol itself does not restrict format, content or length. The (FCS) field contains 16 bits for cyclic redundancy checking (CRC) used in frame checking. They represent the complement of a polynomial remainder calculated on the basis of the (A), (C) and (I) fields.[68]

---

[68] IBM Synchronous Data Link Control--General Information, GA27-3093-1, May 1975, pp. 2-5 to 2-9.

Primary advantage of SDLC with respect to the more familiar BSC procedures is that it exhibits code structure independence. As a bit-oriented protocol, operational requirements for peripheral device control and communications channel control are clearly separated. Many functions are performed by bit designation and manipulation. The more traditional protocols required specific control characters, which sometimes tended to have multiple interpretations.

From a data reliability standpoint, FCS requirements encompass multiple types of fields, such as address, control, and information data, while BSC error control capabilities are limited to protection of text data only. A major weakness of BSC is that the response (DLE Ø or 1) must be transmitted totally unprotected. This can create unrecoverable situations in the event the response is damaged during transmission.[69]

One feature of SDLC not possible with BSC is the ability to initiate and maintain full-duplex data transmission. Assuming the availability of a properly configured (4-wire) communications channel, simultaneous data transmission is possible between two data communications devices. This data flow, coupled with the basic transparency of the protocol with

---

[69] John E. Buckley, "IBM Protocols Part 2: SDLC," Computer Design (February, 1975), pp. 14-16.

respect to code structure, suggests a meaningful improvement over the more restrictive BSC procedures.

Actually, in the age of the microprocessor, talk of "built-in biases" seems a bit archaic.[70] Microcomputer converters have been developed for such things as code, protocol and data rate conversion. By applying such devices to interface problems and, for example, using it for SDLC to BSC conversion, the advantages of new protocols or transparency can be obtained.[71]

b. Commission Findings

On November 25, 1974, oral arguments were held. FCC Chairman Richard E. Wiley queried the participants opposing IBM as to what conditions might be appropriate for IBM's entry as a common carrier. Most replies indicated that their objections would be softened if the Commission imposed a number of conditions on entry. Nicholas Katzenbach, IBM's General Counsel, addressed the "split the market" allegation:

> There is no way we will avoid competing with AT&T. We would have to be absolutely out of our minds--with the Justice Department, Federal Trade Commission and FCC

---

[70] As George Potts, an analyst at Dean Witter and Company noted, "Let's face it, AT&T sets its own standards too." See "Heard on the Street", Wall Street Journal (October 25, 1974).

[71] John E. Buckley, "Microcomputer Converter," Computer Design (February, 1976), pp. 12-14.

looking over our shoulders--to
do so.[72]

The Commission found that the prospect of IBM's entry held promise for bringing to the public new and better specialized communications services, thereby promoting potential realization of DOMSAT policy objectives. IBM's economic and technological capabilities, as well as its extensive experience in the computer and data processing equipment industries, were seen as strong points. Its knowledge of the needs of the business customer was also considered to be of benefit in the development of new and better data communications services.[73]

The Commission concluded that, among other things, the IBM entry would create the potential for a new and different system that was likely to be a strong and vigorous competitor with AT&T in both the domestic satellite and specialized services markets.[74] However, it rejected the proposed corporate ownership structure and ruled that it would entertain any of the following applications:

(a) "Independent Entry Option."

IBM and COMSAT to enter independently of each other; COMSAT to have the option of joining another consortium,

---

[72]"Questions Suggesting IBM's Entry," Telecommunications Reports, 40, No. 48 (December 2, 1974), pp. 1-2.

[73]51 FCC 2d 26-27.

[74]51 FCC 2d 35.

(b) "Balanced CML Option."

> COMSAT and IBM to merge but only with another partner(s), such that no participant would have greater than 49% or less than 10% stock ownership.

(c) "Lease Option."

> COMSAT to provide IBM a space segment and to function only as a carrier's carrier, unable to offer common carrier services directly to the public.[75]

Commissioner Washburn noted in a concurring statement:

> In my judgment the strong technology base of IBM and COMSAT, in their respective fields, promised an unusual degree of public service benefit, should the Commission permit them to proceed as proposed.[76]

In a separate statement, Commissioner Glen O. Robinson also endoresed the decision, believing it to be an affirmative implementation of the basic DOMSAT decision.[77]

But the issue was far from dead. Even as IBM was raising doubt that it would enter the business under the FCC's conditions, five companies[78] took the Commission's decision to the U.S. Court of Appeals

---

[75]William Wewer, "IBM, CML Satellite and the Clayton Act: The FCC's Retreat from Competition in the Telecommunications Industry," Law and Computer Technology (March/April, 1975), p. 34.

[76]51 FCC 2d 46.

[77]Ibid.

[78]RCA Global Communications, RCA Alaska Communications, American Satellite Corp., Fairchild Industries, and the Western Union Telegraph Company.

for the District of Columbia, each of the parties filing a notice of appeal and a petition for review.[79] However, the Court dismissed the cases in February of this year finding "that the order sought to be reviewed in these cases was not final agency action ripe for review [no authorized filing of a restructured CML had been approved by this docket]." But the court did warn that

> our ruling should not be taken as an indication of the court's position with respect to the Commission's handling and disposition of the antitrust issue, particularly as to the need for an evidentiary hearing to determine the competitive impact of the Commission's interlocutory ruling.[80]

3. The SBS Filing Today

On December 22, 1975, Satellite Business Systems, the restructured CML (which is owned 42 1/2 percent each by IBM and COMSAT and 15 percent by the Aetna Casualty and Surety Company), submitted its five volume DOMSAT filing to the Commission for its approval. A systems design aimed primarily at providing direct point-to-point, private line circuit replacement had been dismissed early in the study phase (for the SBS System)

---

[79] "Five Companies Go to Court," Telecommunications Reports, 41, No. 11 (March 17, 1975), p. 5.

[80] The United States Court of Appeals for the District of Columbia Circuit, Order, Nos. 75-1236 through 75-1242, Before: Wright, Tamm and Leventhal, Filed February 23, 1976, George A. Fisher, Clerk.

as not being cost competitive in a market area where other DOMSAT companies had already announced plans to offer services. A system providing private line switched networks (in contrast to point-to-point services only) with integrated voice, data and image capability using TDMA with demand assignment of satellite capacity was chosen instead, having been determined to have both high growth potential and near-term demand.[81] Eight operational transponder channels, each of 54 MHz bandwidth, will have an RF power output of at least 20 watts. Customers will purchase capacity to fit their particular needs, but will also be able to call from a data pool, on demand, additional bandwidth in 224 kilobit increments.

Because SBS anticipates that the initial users of its service will be larger enterprises with relatively large communications requirements, it has examined the five hundred largest United States industrial corporations, as listed by Fortune magazine, and the top 50 companies in each of six additional Fortune categories. From this initial list of 800 companies, SBS selected the 415 companies that appeared to be the more likely customers for the initial type of service that SBS proposed to offer. Demand projections were developed premised upon the

---

[81]Kittiver and Zitzman, "The SBS System," p. 1.

requirements of these companies, to obtain a conservative approximation of the potential demand for the proposed network services. The selection criteria included company revenues, number of employees; present communications usage; installed information processing capability; and geographic dispersion.[82]

Next, in-depth surveys of 16 of these companies, which were believed to be representative of major customer categories, were conducted. This information was then correlated with data obtained from separate sources in order to develop demand projections over a seven year period beginning in 1979. In making these projections, a growth rate of 8 percent per year for voice traffic and 19 percent per year for data traffic was used in obtaining final demand projections for the total number of Earth stations and for total satellite capacity. SBS foresees that a requirement for as many as 7500 such Earth stations could exist by 1986. Using standard interfaces and a totally transparent system, it plans to offer data services from 600 bits per second to 6.312 megabits per second as well as voice and image capabilities to 375 SBS Earth stations.[83]

a. <u>The Bell Bill</u>

Now that the SBS plan was "out-of-the bag,"

---

[82] <u>Amendment to Operational System Applications</u>, Satellite Business Systems, April 16, 1976, pp. 14-16.

[83] Ibid., pp. 24-25. Based on satellite capacity.

AT&T was less guarded about their plans. As a matter of fact they were becoming quite outspoken.

Speaking in New York in January, AT&T's deButts had these corporate comments to share relative to the competitive policies of the Commission:

> We [at AT&T] believe [that]. . . regulated competition. . .is adverse to the interest of the public.
>
> But it is not for us to decide the public interest but rather the public itself. . .it is for this reason that the telephone industry has decided to seek a resolution of the issues confronting it in the only forum I have long felt had the necessary perspective to resolve them--the Congress of the United States. . .
>
> I do not know whether Congress will act or how it will act on the legislation that the telephone industry--with the support of unions representing its employees and the support of many regulators--will shortly seek to have introduced.[84]

It was a bold bid to stifle competitors. DeButts was asking Congress to pass legislation that would stop competition in long-distance services, and permit AT&T or other traditional carriers to acquire the companies that would be put out of business. AT&T has suffered a long series of reversals at the hands of the Commission (which have been upheld in Federal courts), and by persuading Congress to change the rules, it hopes

[84] John D. deButts, "Communications and Public Policy," Presentation before the Fordham Forum (New York: January 28, 1976).

to disarm both the Justice Department (in their pending antitrust actions against AT&T) and the FCC.[85] Some believe that the giant monopoly was not truly concerned with the competitive policies of the Commission initially, but that its perception of SBS (with its potential market penetration) rapidly changed that attitude. The basic bills, H.R. 12323 and S. 3192, were introduced into the Congress on March 4 and March 23 respectively. Since that time, according to Congressional staff personnel, over 100 versions of the bill have been introduced into Congress by over 150 sponsors; it seems there is more than a few opinions of what the public interest is. Hearings are not anticipated to begin this Congress and new legislation will have to be reintroduced next session.

Philip N. Whittaker, president of Satellite Business Systems, commenting on the situation, said:

> Satellite Business Systems has given the preliminary review to the bills recently introduced under the impetus of the monopoly carriers. These bills, if enacted, would appear to make it

---

[85]"AT&T's Bold Bid to Stifle Competitors," Business Week, March 15, 1976, p. 82. Its most recent setback is in the area of "full competition over the entire range of private line services" for specialized carriers. On July 6, 1976 a three judge panel of the U.S. Court of Appeals for the District of Columbia unanimously affirmed the FCC's policy by upholding the grant of specialized common carrier applications to United States Transmission Systems (USTS), turning down a "narrow" interpretation by AT&T which differentiated between services on the basis of present availability.

> virtually impossible for any specialized carrier to compete with the existing monopoly communications carriers, unless the FCC should find that services proposed by a specialized carrier are 'not like or similar' to any services provided by a telephone carrier such as AT&T. This, at best, would require a difficult and protracted proceeding before the FCC.
>
> Quite obviously, this legislative effort seeks to preclude any meaningful competition, and is in direct conflict with the pro-competitive policies supported by the FCC, the White House, the Justice Department and most recently by Rep. Macdonald, Chairman of the House Communications Subcommittee.[86]

The new industries appear to be game for the fight. On May 19, 1976 a group of non-monopoly telecommunications companies announced the launching of their joint effort to oppose the proposed legislation which would eliminate all competition from the telecommunications market. The group is called the Ad Hoc Committee for Competitive Telecommunications (ACCT). MCI's Chairman William McGowan, acting as the spokesman for the group[87] told the press:

> The sole purpose of ACCT is to insure that the Congress--and the consumers of communications--scrutinize this

---

[86]"AT&T, Competitors Square Off Over Bill to Repeal Carterfone," Electronic News (April 12, 1976), p. 50. The Office of Telecommunications Policy has recently released a position paper which opposes the proposed legislation entitled "Competition in Telecommunications--The Telephone Industry Bill," June, 1976.

[87]Datran, Graphnet, Southern Pacific Communications and USTS comprise the group. Although SBS and Telenet are not members of the group, both are reportedly contributing to it financially.

legislation closely and carefully.[88]

b. Potential Competition for AT&T

The SBS DOMSAT system, which is not expected to be in operation until 1979, is called by some "the first major challenge to American Telephone and Telegraph's dominance of domestic communications."[89] Although the direct impact of this application is some years away, 1979 marks the potential convergence of several events that are worthy of note. These include:

(1) SBS's satellite services will commence.

(2) IBM's prohibition from entering the data services market will end.

(3) IBM's next generation of new technology mainframes will most likely be introduced.[90]

These events have been surfaced in articles forewarning of the potential threat of IBM, but there is one more item that should also be included:

(4) the limitation on AT&T's use of satellites to its non-competitive services will be removed.

With the inauguration of the services of AT&T's "COMSTAR" satellite on the Bicentennial, the three-year clock

---

[88] ACCT Press Release, Washington, D. C., May 19, 1976.

[89] "Ma Bell vs. IBM?", Barron's (February 9, 1976), p. 3.

[90] "Market Report-Satellite Business Systems," Telecommunications (June, 1976), p. 16.

on AT&T's domestic satellite began and will stop in 1979 unless the Commission acts to extend the limitation.[91]

Petitions to reject the applications of SBS and set the matter for "full evidentiary hearings" or to defer action were submitted to the Commission on June 1, 1976 by nine parties.[92] Common themes ranged from fears of IBM dominance, questions of SBS's compliance with the basic requirements for joint entry laid down by the Commission, and issues of public interest. Taking note of these actions, SBS in a brief statement restated its beliefs that its system will offer significant pro-competitive benefits to communications users and its intent to vigorously pursue its applications.[93] A 311-page response to these comments was filed with the Commission on August 20, 1976 by SBS.

The House Subcommittee on Communications in a recent print has concluded:

> Common carrier regulation in this country is at a critical state. After decades of neglect by federal regulators the FCC began in the mid-1960's to establish the parameters of a meaningful regulatory program.[94]

---

[91]See generally 38 FCC 2d 676-678 and 35 FCC 2d 851-853.

[92]These are American Satellite, the Computer Industry Association, RCA, Western Union, AT&T, GT&E Satellite, Independent Data Communications Manufacturers, RCA Alascom and the State of Hawaii.

[93]"SBS Applications Come Under Strong Attack," Telecommunications Reports, 42, No. 23 (June 7, 1976), pp. 2-4, 27-31.

[94]U.S. Congress, House, Subcommittee on Communications, Agenda for Oversight: Domestic

The report goes on to point out that the decisions made by the Commission to allow competition with the established carriers have focused attention on the inadequacies of FCC regulation (i.e., personnel, policy planning) without affecting ways of improving it, noting:

> It would be unfortunate if the FCC's current policies of competition were to degenerate into a cartelization of the market because the Commission lacked the resources to do otherwise.[95]

It would also be unfortunate if the benefits of DOMSAT and the new technology were not realized for this same reason. Leland Johnson recently addressed the monopoly/competition issue in the following way:

> [A] particular disconcerting aspect of the current controversy is the underlying assumption that whoever benefits from competition keeps the whole amount and whoever loses thereby suffers irrevocably. There is little recognition of the fact that cost savings to one group may be passed on at least in part to another group. In particular, little consideration is given to the possibility that cost reductions to the business community would be passed on to the consumers--including the low income groups whose interests everyone seems to want to protect.
>
> . . .using rudimentary tools of economic theory, one can show that even a monopolist will pass a portion of any cost reduction to its customers.[96]

Common Carrier Regulations, Subcommittee Print, April 26, 1976, 94th Congress, 2nd Session, p. 26.

[95]Ibid., p. 27.

[96]Leland L. Johnson, "Problems of Regulating

. . . .

At the present time domestic satellite policy is being strongly influenced by the confrontation between the advocates of monopoly on the one hand and the advocates of competition on the other. It appears that the question before the Commission is not "if" SBS will be permitted to compete in the industry but "how" it will enter.

Technological trends, once again, and market potentials have influenced these advocates and encouraged IBM to become a major AT&T competitor and both companies can be expected to structure their products and services to enhance their ultimate competitive postures. When the confrontation does occur many data communications users will be affected as will the public either directly or indirectly. There is insufficient evidence to support some contentions that these effects will be adverse. It is more than likely that the benefits that develop as a result of this competition for service will far outweigh any disadvantages recognized.

With an understanding of the background of these developments, the public can hopfully influence future FCC policy in an advantageous manner. AT&T has been generating much "public impact propaganda"

Specialized Telecommunications Common Carriers," The Rand Paper Series P-5638, May 1976, pp. 8-9.

which should be recognized as such. Information on any of the issues is available from the FCC from the "referee's" point of view. When it comes to the choice between competition and monopoly, their policies have been generally deep rooted and consistent. Even Dean Burch, the conservative Republican, and Nicholas Johnson, perhaps the most liberal Democrat ever to occupy a commissioner's office, voted alike when it came to questions of competition. Although he demeans the Commission, Kinsley also supports competition, noting in the conclusions of his book that without it incentive for innovation disappears and with that goes technological progress.[97]

The Arthur D. Little Company has predicted a world market of telecommunications equipment alone of at least $40 billion a year by 1980 as compared to today's total of only $15 billion a year. Technology will exert pressure for change but it will always be limited by concerns of the marketplace and public policy.[98]

In a 1972 Scientific American dedicated to the theme of communications, Professor Hiroshi Inose of the University of Tokyo wrote:

---

[97]Kinsley, Outer Space and Inner Sanctums, p. 244.

[98]John M. Richardson, "The Technology-Driven Future, the Public Policy-Driven Future or the Market-Driven Future," Communications News (February, 1976), p. 19.

> It is conceivable that a new communications network will be built in which the transmission, switching and processing of speech, data and other information are all performed digitally. Progress in large-scale integrated-circuit technology is drastically cutting the cost of digital hardware.[99]

Such a system will most likely be first realized by Satellite Business Systems.

---

[99] Hiroshi Inose, "Communication Networks," Scientific American, 227, No. 3 (September, 1972), p. 128.

[illegible]

[illegible]

CONCLUSIONS

Modern communications technology, enhanced by the computer, its digital technology, and the advent of the satellite, offers unmatched potential for enhancing this Nation's telecommunications systems and promoting innovative services for the public. The domestic use of communication satellites gives a new dimension to the systems and services of today. But this will happen only if the Nation's telecommunications industry is guided by relevant laws which are contained within a framework of technical and economic order. The laws in turn must stem from a coherent national communications policy which is based on balanced perspectives of the future that lie somewhere between academic optimism, industrial opportunism, and public pessimism.

The regulatory policies of the Federal Communications Commission, particularly over the past ten years, have shown an unprecedented responsiveness to the demands of the public interest and the dynamic technological environment. *Domestic Satellite* is without a doubt one of the finest examples of such progressive policy making to date. Nicholas Zapple has noted that such policies have permitted the regulated communications industry to attain rates of growth and levels of prosperity that surpass all other regulated industries.

The Nation's telecommunication industry will continue to grow and change for the better as a result of the Commission's competitive policies and the opportunities and market forces that they generate. There will always be a need for competition to accommodate the give and take between the availability of various communications capabilities and the applications that people think of for using those capabilities. But as Clay Whitehead has noted:

> ...the ATT "consumer protection" bill shows that the fight for competitive and innovative electronic communications in the U.S. is far from won.[1]

In a recent address to the International Communications Association, FCC Chairman Wiley noted that today's specialized common carriers account for only one-tenth of one percent of the market while AT&T's <u>increase</u> in revenues in the competitive services area in 1975 was more than double the total revenues of all competing suppliers of these services.[2]

---

[1]Letter from Clay T. Whitehead, Santa Monica, California to Robert S. Magnant, October 21, 1976.

[2]Richard E. Wiley, Address before the International Communications Association, 29th Annual Conference (Washington, D.C.: May 3, 1976), pp. 2-3. Total revenues of the specialized carriers do not exceed $170 million while the telephone industires revenues are over $4 billion in the competitive market sectors alone.

Professor Joseph Weizenbaum of MIT has raised this thought in his new book, Computer Power and Human Reason concerning the relationship between policy and new technologies:

> Decisions made by the general public about emergent technologies depend much more on what the public attributes to such technologies than on what they actually are or can and cannot do. If, as it appears to be the case, the public's attributions are wildly misconceived, then public decisions are bound to be misguided and often wrong.[3]

The fact that the Commission has encountered difficulties and opposition to its efforts and that the public is generally unaware of the far-reaching implications of such efforts diminishes neither the laws that guide it nor the agency itself for, in the era of space technology, it has taken in DOMSAT a regulatory giant step.

---

[3]Joseph Weizenbaum, Computer Power and Human Reason (San Francisco, California: W. H. Freeman and Company, 1976), pp. 7-8.

SELECTED BIBLIOGRAPHY

Books

Abramson, Norman, and Kuo, Franklin F., eds. Computer-Communication Networks. Englewood Cliffs, New Jersey: Prentice-Hall, Inc., 1973.

Alderson, W.; Terpstra, V.; and Shapiro, J., eds. Patents and Progress: The Sources of Impact of Advancing Technology. Homewood, Illinois: Irwin Publishing, 1965.

Alexandrowicz, Charles H. The Law of Global Communications. New York: Columbia University Press, 1971.

Borchardt, Kurt. Structure and Performance of the U.S. Communications Industry. Boston: Harvard University Press, 1970.

Clark, Arthur. Voices from the Sky. New York: Harper and Row, 1965.

Codding, George A., Jr. Broadcasting Without Barriers. Paris: UNESCO, 1959.

__________. The International Telecommunication Union. Leiden: E. J. Brill, 1952.

English, H. Edward, ed. Telecommunications for Canada: An Interface of Business and Government. Toronto: Methuen Publishing, 1973.

Feldman, N. E., and Kelly, C. M., eds. Communication Satellites for the 70's. Cambridge, Massachusetts: MIT Press, 1971.

Galloway, Jonathan F. The Politics and Technology of Satellite Communications. Lexington, Massachusetts: D.C. Heath and Company, 1972.

Gillmore, Donald M., and Barron, Jerome A. Mass Communication Law. 2nd ed. St. Paul, Minnesota: West Publishing Company, 1974.

Green, Paul E., Jr., and Lucky, Robert W., eds. Computer Communications. New York: IEEE Press, 1975.

Halliwell, B. J., ed. Advanced Communication Systems. London: Butterworth, 1974.

Irwin, Manley R. The Telecommunications Industry--Integration vs. Competition. New York: Praeger Publishing, 1971.

Jones, Erin Bain. Earth Satellite Telecommunications Systems and International Law. Austin, Texas: The Encino Press, 1970.

Kahn, Alfred E. The Economics of Regulation: Principles and Institutions, Vol. I and II. New York: John Wiley and Sons, 1971.

Kinsley, Michael E. Outer Space and Inner Sanctums: Government, Business and Satellite Communications. New York: John Wiley and Sons, 1976.

Krasnow, Erwin G., and Longley, Lawrence D. The Politics of Broadcast Regulation. New York: St. Martin's Press, Inc., 1973.

Martin, James, and Norman, Adrian R. D. The Computerized Society. Englewood Cliffs, New Jersey: Prentice-Hall, Inc., 1970.

Martin, James. Future Developments in Telecommunications. Englewood Cliffs, New Jersey: Prentice-Hall, Inc., 1971.

Mathison, Stuart L., and Walker, Philip M. Computers and Telecommunications: Issues in Public Policy. Englewood Cliffs, New Jersey: Prentice-Hall, Inc., 1970.

McDonald, John. The Game of Business. Garden City, New York: Doubleday, 1975.

Musolf, Lloyd D. Communications Satellites in Political Orbit. San Francisco: Chandler Publishing Co., 1968.

Peabody, Robert L.; Berry, Jeffrey M.; Frasure, William G.; and Goldman, Jerry. To Enact a Law. New York: Praeger Press, 1972.

Phillips, Charles F., Jr., ed. Competition and Monopoly in the Domestic Telecommunications Industry. Lexington, Virginia: Washington and Lee University, 1974.

Sackman, H., and Nie, Norman, eds. The Information Utility and Social Change. Montvale, New Jersey: AFIPS Press, 1970.

Shepherd, William G., and Gies, Thomas G., eds. Regulation in Further Perspective. Cambridge, Massachusetts: Ballinger Publishing Co., 1974.

Trebing, Harry M., ed. Performance Under Regulation. Lansing, Michigan: MSU Public Utilities Studies, 1968.

White, Eston T. Utilities: Electricity, Gas, Telecommunications. Washington, D.C.: Industrial College of the Armed Forces, 1972.

Winkler, Stanley, ed. Computer Communication--Impacts and Implications. ACM/ICCC, 1972.

Articles

Abramson, Norman. "The ALOHA System." Computer Communications Networks. Englewood Cliffs, New Jersey: Prentice-Hall, Inc., 1973.

__________. "The ALOHA System--Another Alternative for Computer Communications." AFIPS, 37 (1970).

__________. "Packet Switching with Satellites." National Computer Conference, 1973.

Abramson, Norman, and Cacciamani, Eugene R., Jr. "Satellites: Not Just a Big Cable in the Sky." IEEE Spectrum (September, 1975).

Acheson, David C. "Domestic Satellite Developments." Public Utilities Fortnightly, 86 (September 24, 1970).

__________. "Domestic Satellite Proceeding--Status Report." Public Utilities Fortnightly, 90 (December 7, 1972).

Averch, Harvey, and Johnson, Leland L. "Behavior of the Firm Under Regulatory Constraint." American Economic Law Review, 52 (1962).

Binder, R.; Abramson, N.; Kuo, F.; Okinawa, A.; and Wax, D. "ALOHA Packet Broadcasting--A Restrospect." National Computer Conference, 1975.

Brunn, Roy. "Packet Switching and Satellite Technology." *Infosystems*, *21* (August, 1974).

Burch, Dean. "Public Utility Regulation: In Pursuit of the Public Interest." *Public Utilities Fortnightly* (September, 1973).

Cacciamani, Eugene R., Jr. "The SPADE System as Applied to Data Communications and Small Earth Station Operation." *COMSAT Technical Review*, *1*, No. 1 (Fall, 1971).

Coase, R. H. "The Federal Communications Commission." *The Journal of Law and Economics*, *2* (October, 1959).

Cohn, Marcus. "The Federal communications Commission."
Computer
Rutgers Journal of Computers and Law, *1* (Fall, 1970).

Dill, George D. and Shimasaki, Nobuhiko. "Signalling and Switching for Demand Assignment Satellite Communications." *Proceedings of the INTELSAT/IEEE Conference on Digital Satellite Communications*, 1969.

Dingell, John D. "The Role of Spectrum Allocation in Monopoly or Competition in Communications." *Antitrust Bulletin*, *13* (1968).

Dorff, Ervin K. "Computers and Communications: Complementing Technologies." *Computers and Automation* (May, 1969).

Gabel, Richard. "The Early Competitive Era in Telephone Communication, 1893-1920." *Law and Contemporary Problems*, *34*. Durham, North Carolina: Duke University, 1969.

Galloway, George B. "The Operation of the Legislative Reorganization Act of 1946." *American Political Science Review*, *45* (1951).

Geller, Henry. "Competition and Monopoly Policies in Domestic Satellite Communications." *The Antitrust Bulletin*, *13* (Fall, 1968).

Hart, Philip A. "The Congressional Perspective of Competition in the Communications Industries." *The Antitrust Bulletin*, *13* (Fall, 1968).

Heart, F. E.; Kalin, R. E.; Ornstein, S. M.; Crowther, W. R.; and Walden, D. C. "The Interface Message Processor for the ARPA Computer Network." 1972 SJCC: AFIPS Conference Proceedings, 40 (1972).

Horowitz, Andrew R., and Cowlan, Bert. "Should People Fight for Satellites." Tele-VISIONS (January/ February, 1976).

Hough, Roger W. "Future Data Traffic Volume." Computer (September/October, 1970).

Hurley, Neil P. "Satellite Communications." America. 115 (August 27, 1966).

Johnson, Leland L. "Technological Advance and Market Structure in Domestic Telecommunications." American Economic Association, 60 (May, 1970).

Johnson, Nicholas. "Harnessing Revolution: The Role of Regulation and Competition for the Communications Industries of Tomorrow." The Antitrust Bulletin, 13 (Fall, 1968).

__________. "Why Ma Bell Still Believes in Santa." Saturday Review, March 11, 1972.

Kamien, Morton I., and Schwartz, Nancy L. "Market Structure and Innovation: A Survey." Journal of Economic Literature, 13 (March, 1975).

Kestenbaum, Lionel. "The Limits of a Regulated Monopoly: Telephone Attachments, Interconnections, and Use of Circuits." The Antitrust Bulletin, 13 (1969).

Kleinrock, Leonard, and Lam, Simon S. "Packet Switching in a Slotted Satellite Channel." Proceedings, National Computer Conference, 1973.

Kolb, Burton A. "The Rise and Fall of Public Utilities--An Appraisal of Risk." The Journal of Business, 37 (1964).

Legislative Note. "The Communications Satellite Act of 1962." Harvard Law Review, 76 (1962).

Lessing, Lawrence. "Cinderella in the Sky." Fortune, 76 (October, 1967).

Levin, Harvey J. "Organization and Control of Communications Satellites." University of Pennsylvania Law Review, 113 (January, 1965).

Loevinger, Lee. "Regulation and Competition as Alternatives." The Antitrust Bulletin, 11 (1966).

Mathison, Stuart L., and Walker, Philip M. "Regulatory and Economic Issues of Computer Communications." Proceedings of the IEEE, 60 (November, 1972).

__________. "Specialized Common Carriers." Telephone Engineering and Management (October 15, 1971).

McDonald, John. "Getting our Communication Satellite off the Ground." Fortune, 86 (July, 1972).

Morrissey, Michael J., and Smith, John J. "FCC Activities: Domestic Satellites." George Washington Law Review, 41 (May, 1973).

Moulton, Horace P. "Communications Satellites--The Proposed Communications Satellite Act of 1962." Business Lawyer, 18 (November, 1962).

__________. "Monopoly and Competition Issues Facing the Communications Industries." The Antitrust Bulletin, 13 (Fall, 1968).

Murphy, Thomas P. "Federal Regulatory Policy and Communications Satellites: Investing the Social Dividend." The American Journal of Economics and Sociology, 31 (October, 1972).

__________. "Technology and Political Change: The Public Interest Impact of COMSAT." The Review of Politics, 33 (July, 1971).

Nolan, Herbert. "Moving Business Data is Big Business." Rutgers Journal of Computers and Law, 1 (Fall, 1970).

Notes. "The FCC Computer Inquiry: Interfaces of Competitive and Regulated Markets." Michigan Law Review, 71 (November, 1972).

Ornstein, S. M.; Heart, F. E.; Crowther, W. R.; Rising, H. K.; Russell, S. B.; and Mitchell, A. "The Terminal IMP for the ARPA Computer Network." 1972 SJCC: AFIPS Conference Proceedings, 40 (1972).

Phillips, Charles F., Jr. "Domestic Telecommunications Policy: An Overview." Washington and Lee Law Review, 29 (1972).

Podraczky, Emeric. "Utilization of the Geostationary Satellite Orbit." Telecommunications (January, 1975).

Puente, J. G., and Werth, A. M. "Demand-Assignment Service for the INTELSAT Global Network." IEEE Spectrum (January, 1971).

Riegel, O. W. "Communications by Satellite: The Political Barriers." Quarterly Review of Economics and Business, 11 (1971).

Roberts, Lawrence G. "Data by the Packet." IEEE Spectrum (February, 1974).

__________. "Dynamic Allocation of Satellite Capacity Through Packet Reservation." National Computer Conference, 1973.

Sadofsky, Michael. ". . .This Proceeding is Terminated." Telecommunications (January, 1973).

Schiller, Herbert I. "Communications Satellites: A New Institutional Setting." Bulletin of the Atomic Scientists, 23, No. 4 (April, 1967).

Shayon, Robert Lewis. "Bird Watching." Saturday Review, April 17, 1971.

Shepherd, William G. "The Competitive Margin in Communications." Technological Changes in Regulated Industries. Washington, D.C.: The Brookings Institution, 1971.

Silberman, Charles E. "The Little Bird That Casts A Big Shadow." Fortune, 75 (February, 1969).

Smith, Delbert D. "The Interdependence of Computer and Communications Services and Facilities: A Question of Federal Regulation." University of Pennsylvania Law Review, 117 (April, 1969).

Smythe, Dallas W. "The 'Orbital Parking Slot' Syndrome and Radio Frequency Management." Quarterly Review of Economics and Business, 12 (1972).

Stanley, Kenneth B. "International Telecommunications Industry: Interdependence of Market Structure and Performance Under Regulation." Land Economics (November, 1973).

Strassburg, Bernard. "Communications and Computers: How Shall the Twain Meet?" Public Utilities Fortnightly, 82 (September 12, 1968).

________. "Competition and Monopoly in the Computer and Data Transmission Industries." The Antitrust Bulletin, 13 (Fall, 1968).

________. "New Technology and Old Institutions." Telecommunications (June, 1974).

Taub, Barry. "Federal Communications Commission Regulation of Domestic Computer Communications: A Competitive Reformation." Buffalo Law Review, 22 (Spring, 1972).

Trebing, Harry M. "Common Carrier Regulation--The Silent Crisis." Law and Contemporary Problems, 34. Durham, North Carolina: Duke University, Spring, 1969.

Turner, Donald F. "The Scope of Antitrust and Other Economic Regulatory Policies." Harvard Law Review, 82 (1969).

Vermillion, Lois. "Dean Burch: FCC's Pragmatic Boss." Electronics (September 28, 1970).

Walters, Wynn. "Computer Communications: The Start of a Revolution." The Business Quarterly (Winter, 1973).

Werner, Robert L. "A Lawyer Looks at Our Communications Policy." Jurimetrics Journal, 11 (December, 1970).

Werth, Andrew M. "SPADE: A PCM FDMA Demand Assignment System for Satellite Communications." Proc. INTELSAT/IEEE Conference on Digital Satellite Communications, 1969.

Ziegler, Hans K. "Space Communications--A Major Candidate for Commercial Utilization of Space." Advances in the Astronautical Sciences, 28 (1968).

Government Documents

The Communications Act of 1934, with Amendments and Index Thereto. Washington, D.C.: U.S. Government Printing Office.

The Communications Satellite Act of 1962. Pub. L. No. 87-624, 87th Congress, 2nd Session, August 31, 1962, 76 Stat. 419.

U.S. Congress. Message from the President of the United States. House of Representatives, House Document No. 157, 90th Congress, 1st Session, August, 1967.

U.S. Congress, House, Committee on Government Operations. Satellite Communications. H.R. No. 178, 89th Congress, 1st Session, March 17, 1965.

U.S. Congress, House, Committee on Science and Astronautics. Hearings on Communications Satellites. 87th Congress, 1st Session, May 8-10 and July 13, 1961.

__________. Commercial Applications of Space Communications Systems. H.R. No. 1279, 87th Congress, 1st Session, October 11, 1961.

__________. Assessment of Space Communications Technology. H.R. No. 859, 91st Congress, 2nd Session, March 3, 1970.

U.S. Congress, House, Subcommittee on Communications and Power. First Session on the Jurisdiction and Activities of the Federal Communications Commission. 91st Congress, 1st Session, March 6, 1969.

__________. First Session on the Jurisdiction and Activities of the Federal Communications Commission. 92nd Congress, 1st Session, April 29, 1971.

__________. Interim Report and Recommended Courses of Action Resulting from the Hearings on Telecommunications Research and Policy Development. 94th Congress, 2nd Session, December, 1975.

__________. Agenda for Oversight: Domestic Common Carrier Regulation. Committee Print, 94th Congress, 2nd Session, April 26, 1976.

U.S. Congress, Senate, Committee on Aeronautics and Space Sciences. Documents on International Aspects of the Exploration and Use of Outer Space. S. Doc. No. 18, May 9, 1963.

U.S. Congress, Senate, Committee on Commerce. Communications Satellite Act of 1962. S.R. No. 1584, 87th Congress, 2nd Session, June 11, 1962.

__________. Hearings on Communications Satellite Legislation. 87th Congress, 2nd Session, April 10-13, 16, 24, and 26, 1962.

U.S. Congress, Senate, Committee on Interstate Commerce. S.R. No. 781, 73rd Congress, 2nd Session, April 17, 1934.

U.S. Congress, Senate, Subcommittee on Communications. Progress Report on Space Communications. Serial 89-78, 87th Congress, 2nd Session, August 10, 17, 18, and 23, 1966.

__________. Overview of the Federal Communications Commission. 92nd Congress, 2nd Session, February 1 and 8, 1972.

__________. Overview of the Federal Communications Commission. 93rd Congress, 1st Session, February 22, 1973.

U.S. Department of State. Treaties and Other International Acts. Series No. 5646, International Telecommunications Satellite Consortium (INTELSAT), 1964.

U.S. Federal Communications Commission. Allocation of Frequencies in the Bands Above 890 MC. 27 FCC 359 (1959).

__________. Authorized Entities and Authorized Users Under the Communications Satellite Act of 1962. Docket No. 16058, 4 FCC 2d 421 (1966).

__________. "COMSAT May Furnish Satellite Services and Channels Only to Other Common Carriers Except in Unique Circumstances." Public Notice. 4 FCC 2d 12 (1966).

__________. Notice of Inquiry, Establishment of Domestic Noncommon Carrier Communication Satellite Facilities by Nongovernmental Entities. Docket No. 16495, 2 FCC 2d 668 (1966).

__________. Notice of Proposed Rulemaking. Docket No. 16495, 22 FCC 2d 810 (1970).

__________. Petition for Approval of Changes in Corporate Structure of CML Satellite Corp. Docket No. 20221, 51 FCC 2d 14.

__________. Regulatory and Policy Problems Presented by the Interdependence of Computer and Communication Services and Facilities. Docket No. 16979, 7 FCC 2d 11 (1966).

__________. Report and Order. Docket No. 16495, 22 FCC 2d 86 (1970).

__________. Second Report and Order. Docket No. 16495, 35 FCC 2d 844 (1972).

__________. Use of the Carterfone Device in Message Toll Telephone Service. Docket No. 16942, 13 FCC 2d 420 (1968).

__________. 38th Annual Report/Fiscal Year 1972. Washington, D.C.: U.S. Government Printing Office.

__________. 39th Annual Report/Fiscal Year 1973. Washington, D.C.: U.S. Government Printing Office.

The White House. Memorandum for the Honorable Dean Burch, Chairman of the Federal Communications Commission. 22 FCC 2d 125 (January 23, 1970).

Periodicals and Newspapers

Aviation Week and Space Technology, 1965-1976. Published weekly by McGraw-Hill Publishing Company, 1221 Avenue of the Americas, New York, New York 10020.

Business Week, 1965-1976. Published weekly by McGraw-Hill Publishing Company, 1221 Avenue of the Americas, New York, New York 10020.

Broadcasting, 1965-1976. The Newsweekly and Broadcasting and Allied Arts. Published by Broadcasting Publications, Inc., 1735 DeSales Street, N.W., Washington, D. C. 20036

Computer Decisions, 1970-1976. Distributed Computing for Management. Published monthly by Hayden Publishing Company, Inc., 50 Essex Street, Rochelle Park, New Jersey 07662.

Computer Design, 1970-1976. The Magazine of Digital Electronics. Published monthly by Computer Design Publishing Corporation, 221 Baker Avenue, Concord, Maine 01742.

Datamation, 1970-1976. A computer industry monthly. Published by Technical Publishing Company, 1301 Grove Avenue, Barrington, Illinois 60010.

Electronics, 1970-1976. The international magazine of Electronics Technology. Published bi-weekly by McGraw-Hill Publishing Company, 1221 Avenue of the Americas, New York, New York 10020.

Electronic News, 1970-1976. Published weekly by Fairchild Publications, Inc., 7 East 12th St., New York, New York 10003.

IEEE Spectrum, 1968-1976. Published monthly by the Institute of Electrical and Electronics Engineers, 345 East 47th Street, New York, New York 10017.

IEEE Transactions on Communications, 1968-1976. A publication of the IEEE Communications Society. Published monthly by the Institute for Electrical and Electronics Engineers, 345 East 47th Street, New York, New York 10017.

The New York Times, 1965-1976. Published daily by the New York Times, 229 West 43rd Street, New York, New York 10036.

Public Utilities Fortnightly, 1965-1976. Published bi-weekly by Public Utilities Reports, Inc., Suite 500, 1828 L Street, N.W., Washington, D.C. 20036.

Scientific American, 1965-1976. Published monthly by Scientific American, Inc., 415 Madison Avenue, New York, New York 10017

Telecommunications, 1970-1976. An international monthly on data communications. Published by Horizon House, 610 Washington Street, Dedham, Massachusetts 02026.

The Telecommunications Journal, 1969-1973. Monthly Magazine of the International Telecommunications Union. CH-1211 Genevre 20, Switzerland.

Telecommunications Reports, 1973-1976. Authoritative newsweekly of the Telecommunications Field, 1204 National Press Building, Washington, D.C. 20004.

Telephony, 1970-1976. The Journal of the Telephone Industry. Published weekly by Telephony Publishing Corporation, 53 West Jackson, Chicago, Illinois 60604.

The Wall Street Journal, 1965-1976. Published daily except Saturdays, Sundays and Legal Holidays, 1701 Page Mill Road, Palo Alto, California 94304.

Miscellaneous Sources

Applications of Computer/Telecommunications Systems, Organization for Economic Cooperation and Development Informantic Studies 8. Paris: OECD, 1975.

Codding, George A., Jr. "The U.S. and the ITU in a Changing World." OT Special Publication 75-6, Department of Commerce, December 1975.

Computers and Telecommunications, Organization for Economic Cooperation and Development Informantics Studies 3. Paris: OECD, 1973.

Fackelman, Mary P., and Krekel, Kimberly A. "International Telecommunications Bibliography." OT Special Publication 76-7, Department of Commerce, March 1976.

Final Report, President's Task Force on Communications Policy, December 7, 1968. Washington, D.C.: U.S. Government Printing Office, 1969.

Gallawa, R. L.; Bloom, L. R.; Hanson, A. G.; Gray, E. M.; and Hull, J. A. Telecommunications Alternatives with Emphasis on Optical Waveguide Systems. U.S. Department of Commerce, OT Report 75-72, October, 1975.

Hatfield, Dale. "A General Analysis of Domestic Satellite Orbit/Spectrum Utilization." U.S. Department of Commerce, Office of Telecommunications, NTIS PB 207397, December 13, 1971.

Hough, R. W.; Fratessa, Carolyn; Holley, Virginia; Samuel, A. H.; and Wells, L. J. A Study of Trends in the Demand for Information Transfer. Menlo Park, California: Stanford Research Institute, 1970.

Hyde, Rosel H. "Space Age Regulation." Presentation before the Midwest Association of Railroad and Utilities Commissioners. Chicago, Illinois: July 7, 1964.

International Telecommunication Union, Joint World Plan Committee of the CCIR and CCITT. General Plan for the Development of Interregional Telecommunication Networks, 1967-1970-1975. Geneva: ITU, 1968.

Johnson, Leland L. "Problems of Regulating Specialized Telecommunications Common Carrier." The Rand Paper Series P-5638, May 1976.

Johnson, Nicholas. "The Capacity to Govern: The Role of the FCC in the Development of National Policy for Computer Communications." Hopkins-Brooking Lecture Series, Computer Communications and the Public Interest, Advanced International Studies, March 12, 1970.

Kittiver, C., and Zitzman, F. R. "The SBS System--An Innovative Domestic Satellite System for Private-Line Networks." Presentation before the AIAA/CASI Sixth Communication Satellite Systems Conference. Montreal, Canada: April 5, 1976.

Laskin, Paul L. Planning for a Planet, An International Discussion on the Structure of Satellite Communications. New York: The Twentieth Century Fund, 1971.

Leming, Tom. "An Appraisal of the Role of Satellites in Domestic Communications." Presentation before the AIAA 4th Communications Satellite Systems Conference. Washington, D.C.: April 14-16, 1972.

Melody, William. "Technological Determinism and Monopoly Power in Communications." Presentation before the American Economic Association. New Orleans, Louisiana: December 28, 1971.

The Network Project. Domestic Communications Satellites. New York: Its Notebook No. 1, 1972.

Report by the Telecommunication Science Panel of the Commerce Technical Advisory Board. "Electromagnetic Spectrum Utilization--The Silent Crisis." Washington, D.C.: U.S. Department of Commerce, October 1966.

Reports on Selected Topics in Telecommunications. The Final Report by the Committee on Telecommunications, National Academy of Engineering. Washington, D.C.: National Academy of Sciences, November, 1968.

Satellite Business Systems. _Application of Satellite Business Systems for a Domestic Communications Satellite System_, Volumes I, II, and III, December 1975.

Selective List of Studies and Reports on Communication Satellite Projects. Paris: UNESCO, December 1973.

Weil, Gordon L. _Communicating by Satellite: An International Discussion_. New York: Twentieth Century Fund, 1969.

Wiley, Richard E. Address Before the International Communications Association, 29th Annual Conference. Washington, D.C.: May 3, 1976.

APPENDICES

APPENDIX A

LETTERS, INTERVIEWS OR DOCUMENTS FURNISHED

Norman Abramson, Director
The ALOHA System
University of Hawaii
Honolulu, Hawaii 96822

Walter Adams
Distinguished University Professor
Michigan State University
East Lansing, Michigan 48823

James H. Alleman, Economist
Office of Telecommunications
U.S. Department of Commerce
Boulder, Colorado 80302

Robert S. Black
Exxon Company, U.S.A.
P. O. Box 2180
Houston, Texas 77001

Louis R. Bloom
Office of Telecommunications
U.S. Department of Commerce
Boulder, Colorado 80302

Robert Bottomley
Satellite Business Systems
1750 K Street, N.W.
Washington, D.C. 20006

Dean Burch, Esq.
Pierson, Ball and Dowd
1200 18th Street, N.W.
Washington, D.C. 20036

Carl J. Cangelosi, Esq.
RCA American Communications, Inc.
60 Broad Street
New York, New York 10004

Abram Chayes, Professor
Harvard Law School
Cambridge, Massachusetts 02138

Alexander J. Chisholm, Assistant Vice President
Regulatory Matters
Western Union
1828 L Street, N.W.
Washington, D.C. 20036

R. H. Coase
Clifton R. Musser Professor of Economics
University of Chicago Law School
1111 East 60th Street
Chicago, Illinois 60637

Ms. Gail Crotts
Consumer Assistance Office
Federal Communications Commission
Washington, D.C. 20554

Anthony Cusumano
Director of Traffic
American Broadcasting Company
1330 Avenue of the Americas
New York, New York 10019

Marvin J. Diamond, Esq.
Hogan and Hartson
815 Connecticut Avenue
Washington, D.C. 20006

Wayne Du Bois
AT&T Long Lines
32 Avenue of the Americas
New York, New York 10013

Nathaniel E. Feldman
The Rand Corporation
Santa Monica, California 90406

Joseph R. Fogarty, Esq.
Communications Counsel
United States Senate
Committee on Commerce
Washington, D.C. 20510

Jonathan F. Galloway, Associate Professor
Department of Politics
Lake Forest College
Lake Forest, Illinois 60045

John F. Gerstner
Satellite Business Systems
1750 K Street, N.W.
Washington, D.C. 20006

Richard P. Gifford, Vice President
Communications Projects
General Electric Company
Lynchburg, Virginia 24502

W. J. Gorman, Jr.
Public Relations Director
American Telephone and Telegraph Company
195 Broadway
New York, New York 10007

R. H. Harvey, Manager
Mathematics, Computers and Systems Department
Exxon Corporation
P. O. Box 153
Florham Park, New Jersey 07932

Dale N. Hatfield, Chief
Office of Plans and Policy
Federal Communications Commission
Washington, D.C. 20554

Fred W. Henck, President and Editor
TELECOMMUNICATIONS REPORTS
1204 National Press Building
Washington, D.C. 20045

Andrew R. Horowitz, Co-Director
Public Interest Satellite Association
55 W. 44th Street
New York, New York 10036

Manley R. Irwin, Professor of Economics
The Whittemore School of Business and Economics
University of New Hampshire
Durham, New Hampshire 03824

Leland Johnson, Director
Communications Policy Program
The Rand Corporation
Santa Monica, California 90406

Nicholas Johnson
P. O. Box 19101
Washington, D.C. 20036

Alfred E. Kahn, Chairman
New York Public Service Commission
Empire State Plaza
Albany, New York 12226

Dr. Jeffrey Krauss
Office of Plans and Policy
Federal Communications Commission
Washington, D.C. 20554

Ms. Pat Kuhlman
Ad Hoc Committee for Competitive
Telecommunications
8130 Boone Boulevard
Vienna, Virginia 22180

Sebastian Lasher
Office of Commissioner Washburn
Federal Communications Commission
Washington, D.C. 20554

Robert W. Lucky, Head
Advanced Data Communications Department
Bell Laboratories
Holmdel, New Jersey 07733

Andrew Margeson, Staff Economist
U.S. House of Representatives
Committee on Interstate and Foreign
Commerce
Washington, D.C. 20515

Stuart G. Meister, Esq.
Regulatory Counsel
American Satellite Corporation
20301 Century Boulevard
Germantown, Maryland 20767

Ralph Metts
International Business Machines
System Communications Division
Gaithersburg, Maryland 20760

Thomas P. Murphy, Professor and Director
Institute for Urban Studies
University of Maryland
College Park, Maryland 20742

Harry Newton
Telecommunications Writer and Consultant
3 Sheridan Square
New York, New York 10014

G. J. Rauchenbach, Director
Congressional and Government Relations
Communications Satellite Corporation
950 L'Eufant Plaza, S.W.
Washington, D.C. 20024

Paul Rodgers, General Counsel
National Association of Regulatory
Utility Commissioners
1102 Interstate Commerce Commission Building
P. O. Box 684
Washington, D.C. 20044

William B. Rogers
Manager of Data Communications
American Broadcasting Company
1330 Avenue of the Americas
New York, New York 10019

William Schmidt, Jr.
Ground Systems Engineering
Satellite Business Systems
1750 K Street, N.W.
Washington, D.C. 20006

William G. Shepherd
Professor of Economics
University of Michigan
Ann Arbor, Michigan 48104

Kenneth R. Stanley, Economist
Common Carrier Bureau
Federal Communications Commission
Washington, D.C. 20554

Richard Talley
230 Via Anita
Redondo Beach, California 90277

Harry M. Trebing, Director
Institute of Public Utilities
Professor of Economics
Michigan State University
East Lansing, Michigan 48824

Philip M. Walker
Vice President and General Counsel
Telenet Communications Corporation
Washington, D.C. 20036

Commissioner Abbott Washburn
Federal Communications Commission
Washington, D.C. 20554

Harry Watson
Executive Office of the President
Council of Economic Advisors
Washington, D.C. 20506

Gordon L. Weil
Political Intelligence, Inc.
Harpswell, Maine 04079

Clay T. Whitehead, President
Allison Technical Services
1520 Arizona Avenue
Santa Monica, California 90404

Albert D. Wheelon
Vice President and Group Executive
Hughes Aircraft Company
Space and Communications Group
P. O. Box 92919
Worldway Postal Center
Los Angeles, California 90009

Craig F. Williamson
Director of Personnel and Administration
Satellite Business Systems
1750 K Street, N.W.
Washington, D.C. 20006

Nicholas Zapple, Esq.
6518 Valley Court
Falls Church, Virginia 22042

Ms. Cheryl Zegers
Technical Information Specialist
Department of Commerce Library
Boulder, Colorado 80303

APPENDIX B

A CHRONOLOGY OF INDIVIDUALS INTERACTING IN DOMSAT

	1965 / 1970 / 1975
The Presidency	John F. Kennedy O Lyndon B. Johnson O Richard M. Nixon O Gerald R. Ford
Senate Commo Subcommittee	Senator John O. Pastore - Chairman from 1955 to 1976 O
House Commo Subcommittee	Congressman Torbert H. MacDonald - Chairman from 1967 to 1976 O
Commission Chairman	Newton Minow O E. William Henry O Rosel H. Hyde O Dean Burch O Richard E. Wiley
Common Carrier Chief	O Bernard Strassburg from 1964 to 1973 O Walter R. Hinchman
Primary DOMSAT Policy Dates	O COMSAT Act of 1962; O ABC's Proposal; O Notice of Inquiry; O Johnson Message; O Task Force Report; O Flanigan Memorandum; O Report And Order; O Second Report; O Final Report; O IBM Entry; O CML Decision; O SBS Filing; O Petitions to Deny

APPENDIX C

FURTHER ANALYSIS OF THE ALOHA CONCEPT

For those interested in the details of the ALOHA concept and its protocol considerations, this supplementary information has been prepared based on information contained in several journal articles prepared by Abramson, Binder, Roberts and others.

Analysis

In Figure 7, "$\underline{t}$" is defined as the duration of the packet (34 milliseconds including receiver synchronization) and we make the pessimistic assumption that when overlap occurs neither packet is received without error. We define a random point process for $\underline{K}$ active users by focusing our attention on the starting times of the packets and make a distinction between packets transmitted for the first time (message packets) and retransmissions (repetitions). Lambda ($\underline{L}$) is the average rate of occurrence of message packets from a single user (all users are assumed to have this identical rate). Thus the random point process from all active users has an average rate of occurrence of

$$\underline{r} = \underline{KL}$$

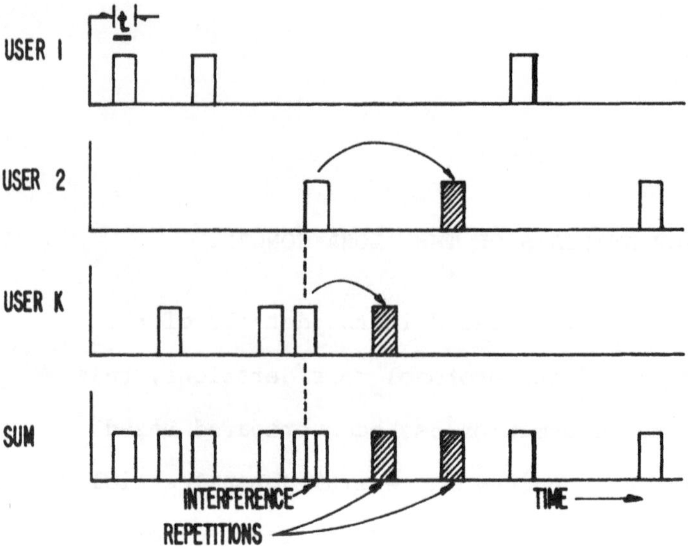

Source: Abramson, "The ALOHA System"

Figure 7. ALOHA Multiplexing

where r is the average number of message packets per unit time from K users. If we were able to pack the messages into the available channel space perfectly with no space between, then rt, the channel utilization, would equal 1. Since the channel utilization is proportional to K users, the objective is to determine the maximum value of K.

First R is defined as the average number of message packets plus repetitions per unit time from K users. R must be greater than r if there are any repetitions. Rt is thus defined as channel traffic. It is now assumed that the inter-arrival times of the point process defined by the start times are independent and exponential. This is reasonable if the retransmission delay is large compared to t and the number of retransmissions is not too large (the ratio of the average time between packets for each user to the single packet transmission time for the ALOHANET is about 2000:1).

Under the exponential assumption, the probability that there will be no starts (of either message packets or repetitions) in time interval T is $\exp(-RT)$. Using this, a packet will overlap with another packet if there exists at least one other start point t or less seconds before or after the start of the given packet. Thus the probability that a given message will be repeated is

$$1 - \exp(-2Rt)$$

If we relate R (the average number of message packets plus repetitions) to r (the average number of message packets) we get

$$R = r + R\,[1 - \exp(-2Rt)]$$

or

$$rt = Rt\,[\exp(-2Rt)]$$

Note that the channel utilization (or the capacity of the random access channel) reaches a maximum value of 1/2(exp) = 0.184, at which point the traffic on the channel becomes unstable and the average number of retransmissions becomes unbounded. The maximum number of interactive users the system can support can be found by setting

$$\underline{rt} = \underline{KLt} = 1/2(\exp)$$

A conservative estimate for $\underline{L}$ is 1/60 corresponding to each active user sending a message packet at an average rate of once a minute. With $\underline{t}$ equal to 34 milliseconds, then

$$K_{MAX} = 324.$$

Bear in mind additionally that while a user is not active he consumes no channel capacity so that the total number of users of the system can be a considerably larger number than the 324.[1]

It was later realized that it is possible to modify the completely unsynchronized use of the ALOHA channel described in order to increase the channel capacity. In pure ALOHA each user simply transmits a packet when ready without any attempt at coordination with other users, leading to somewhat inefficient channel utilization. Slotted ALOHA establishes a time base and each user is required to start his packets only at certain fixed instants. The channel is slotted into segments of time equal to the packet

---

[1]Norman Abramson, "The ALOHA System," Computer Communications Networks (Englewood Cliffs, New Jersey: Prentice Hall, 1973).

transmission time and users are required to begin transmissions at the beginning of time slots. The access is still random in the sense that terminals transmit into random slots, and if necessary retransmit after waiting a random number of slots. In this fashion, if two or more messages conflict they will overlap completely rather than partially. It has been shown that the effective capacity of this scheme is 1/exp of the nominal capacity (twice the previous capacity) using the same assumptions as before.[2]

A third method for using these channels is to attempt to schedule their use in some direct fashion; this introduces the notion of a reservation system in which time slots are reserved for specific users' transmissions. Every M slots, one slot is subdivided into V small slots for reservations and acknowledgments. The remaining M large slots are RESERVED for data packets. A reservation is a request for from one to eight RESERVED slots. Both the ALOHA and the RESERVED technique depend for their efficiency on the total multi-station traffic rather than individual station traffic. The reservation system is a factor of three more efficient than the ALOHA system and for large

---

[2]Norman Abramson, "Packet Switching with Satellites," National Computer Conference, 1973. Figure 8 is a plot of Rt versus rt, for a pure ALOHA channel and a slotted ALOHA channel.

Source: Abramson, "Packet Switching with Satellites"

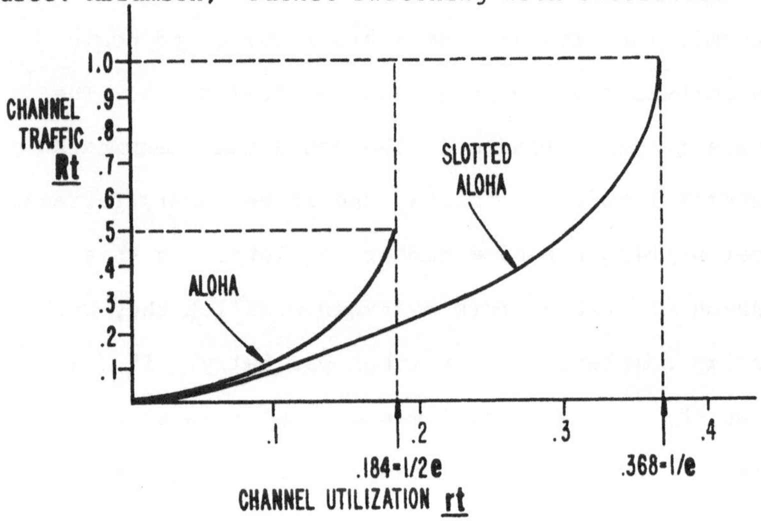

Figure 8. Channel Utilization vs. Channel Traffic

(100 KB) traffic levels achieves almost perfect channel utilization.[3]

Protocol Considerations

Several of the original choices for the ALOHA concept have undergone significant changes as a result of new resources and interfaces or advancements in theoretical knowledge. Binder, Abramson, Kuo, Okinawa, and Wax recently updated several aspected of the concept.

(1) Random access control. The retransmission strategy used in the random access scheme plays a central role in the scheme's effectiveness. It can

---

[3]Lawrence G. Roberts, "Dynamic Allocation of Satellite Capacity through Packet Reservation," National Computer Conference, 1973.

affect the average delay experienced by users and prevent channel saturation (where the channel becomes filled with retransmissions and the number of successful packets falls to zero). These topics have only been recently quantified in the dissertations of Metcalf at MIT and Lam at UCLA. One approach is to use different retransmission constants at each node to avoid packet conflicts. This results in a priority structure. By simply dividing users into two groups--one transmitting at high power and the other at low power--Metzner notes that another type of hierarchical structure can be established (increasing channel utilization to about 53 percent in the process). Programmable Control Units (PCU's) are another way in which a high degree of systems protocol flexibility can be permitted.[4]

(2) Channel queuing. The MENEHUNE acts as a concentrator for the broadcast channel, queuing waiting traffic when necessary for sequential transmission to user nodes. It is important that the return channel traffic not prevent the prompt return of an ACK to a user and cause unnecessary repetitions. A scheme with an ACK queue, as shown in Figure 9, is

---

[4] R. Binder, N. Abramson, F. Kuo, A. Okinawa, and D. Wax, "ALOHA Packet Broadcasting--A Retrospect," National Computer Conference Proceedings, 1975, p. 205. see also J. J. Metzner, On Improving Utilization in ALOHA Networks, IEEE Transactions on Communications, April 1976, pp. 447-448.

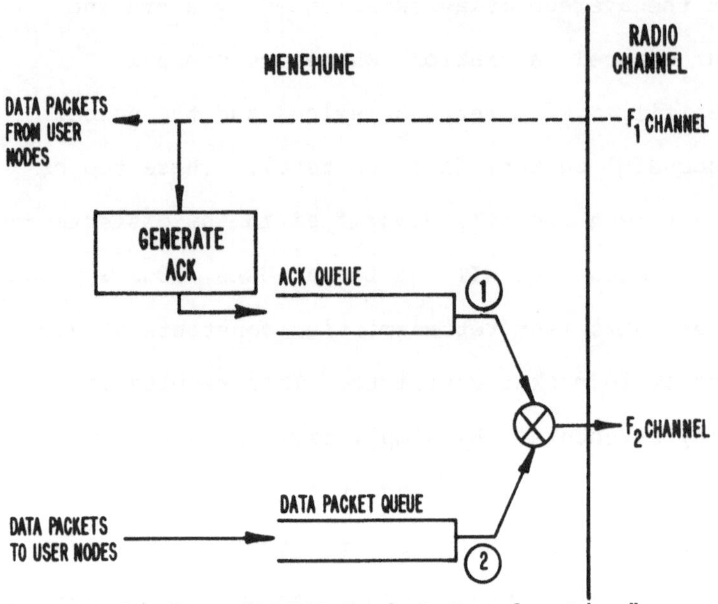

Source: Binder, "ALOHA Packet Broadcasting"

Figure 9. Priority Queues

used so that the ACK's are checked for after each F_2 channel transmission. This guarantees that at most one complete data packet plus any previously queued ACK's will be sent ahead of the next data packet. Hogging of the channel by a few users is also eliminated by queuing. Only one packet per user is allowed on the queue at any one time.[5]

(3) Packet length. The initial use of "half-packets" proved to be a significant source of hardware and software bugs and defeated the "simplified design and construction of the hardware" purpose of fixed-length packets. An end-of-line (EOL--eight zero

---

[5]Ibid., pp. 206-207.

bits) indicator was used to identify the end of actual data, which was restricted to 7-bit ASCII. Formats have now been changed to allow for variable length packets. An 8-bit count field is used in the header to indicate the number of 8-bit data bytes in the packet. This eliminates wasted channel capacity and the unambiguous detection of the EOL indicator.[6]

(4) Error control. Because of the high probability of errors at full loading of the random access channel, reliable error detection is achieved with two 16-bit parity check words, one following the header and one following the data. The separate header check forms the basis for reliable packet synchronization (discussed below) and accurate establishment of the packet length and other information prior to processing of the packet data.[7]

(5) Flow control. Complications to the MENEHUNE flow control processing were caused by the connection of the ALOHANET to the ARPANET via a 50 kilobit INTELSAT IV satellite path. The quarter second propagation time and the ARPANET protocol required a substantial increase in the size of the MENEHUNE buffer pool and a more complicated queuing structure. To support broadcast channels, generally

---

[6] Ibid., p. 207.

[7] Ibid., p. 208.

more than one packet per user must be stored. A large amount of buffering is required at the receive end of the link to support continuous display at higher speed terminals. A 9600 bps terminal requires approximately a 1000-byte buffer. In general CRT terminal users do not require continuous output at this rate and a smaller amount of buffering is in fact used. To maintain the single-packet-per-user policy for the channel queue, a separate queue was created for each user to hold additional packets. The maximum allowed size for each user queue is adaptive to the user's terminal rate and buffer pool availability.[8]

Broadcast channel flow control was necessary since each line (packet) sent to a user node must be completely displayed before a new line can be received. This was accomplished by the scheme shown in Figure 10, in which the control for each user node is centralized at the MENEHUNE. The latter counts off the required display time following transmission of each packet to a user, inhibiting further transmission to that user until the time is up. To prevent 360 output from tying up MENEHUNE buffers while packets are being displayed, a handshaking flow control is used; the 360 sends only one line of output for each user, then waits for a go-ahead (GA) message with that user's

---

[8] Ibid., p. 210.

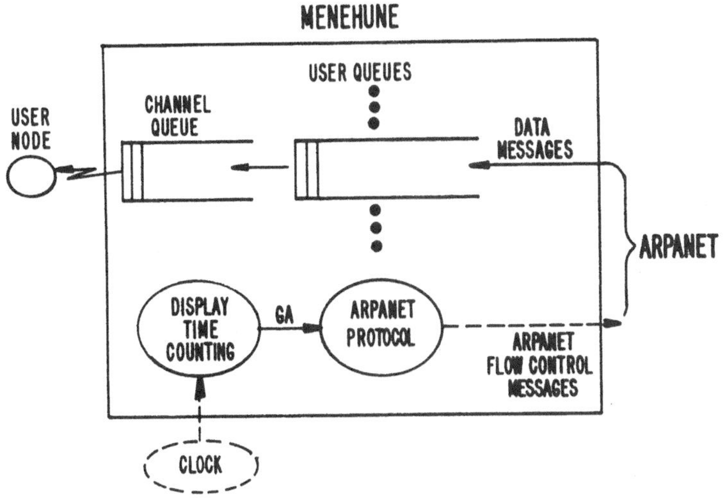

Source: Binder, "ALOHA Packet Broadcasting"

Figure 10. Broadcast Channel Flow Control

address. The GA is sent by the MENEHUNE whenever a user's display time is up, resulting in at most one buffer required for each user. Many ARPANET computers interact with their users full duplex. Since no explicit flow control was normally provided for input from users to the MENEHUNE, users were forced to either interact in a half-duplex mode or suffer occasional losses of input data. Added buffer pool capability helps permit full duplex interaction. If the ARPANET slows down, an excessive amount of buffers could become queued in the MENEHUNE on behalf of the user. To prevent buffer hogging, a count of input queues per user is maintained and when maximums are reached, arriving packets are discarded and the user is so notified.[9]

[9]Ibid., p. 211.

(6) Synchronization. ALOHANET requires special synchronization techniques in the modem and data terminal equipment. The phase-shift-keying used in the modem design is a bit-synchronous technique and bit synchronization must be performed in the demodulator before packet synchronization can be attempted. The bit-sync detection is designed to provide very low false detection (less than 10^{-6}) and high probability of packet detection. This requires a bit-sync preamble of 90 bits for reliability. By redesigning the phase-lock circuit, this can be reduced to about 10 bits. Present plans are to do away with the bit-sync preamble entirely. Packet synchronization in the data terminal buffer is accomplished by the 16-bit parity word in the packet header. When the parity check routine accepts the header, the packet is assumed to be synchronized. Less than one out of a thousand packets will be lost due to packet sync errors.[10]

[10] Ibid., pp. 212-213.